MPI Series in
Biological Cybernetics

Vol. 52

MPI Series in Biological Cybernetics

Vol. 52

Editor:

Prof. Dr. Heinrich H. Bülthoff

Max Planck Institute for Biological Cybernetics
Max-Planck-Ring 8
D-72076 Tübingen
Germany

Sujit Rajappa

Towards Human-UAV Physical Interaction and Fully Actuated Aerial Vehicles

Logos Verlag Berlin

Bibliographic information published by the Deutsche Nationalbibliothek

The Deutsche Nationalbibliothek lists this publication in the Deutsche Nationalbibliografie; detailed bibliographic data are available on the Internet at http://dnb.d-nb.de .

ISBN 978-3-8325-4767-7
ISSN 1618-3037

Logos Verlag Berlin GmbH
Comeniushof, Gubener Str. 47,
D-10243 Berlin, Germany

phone: +49 (0)30 / 42 85 10 90
fax: +49 (0)30 / 42 85 10 92
http://www.logos-verlag.com

Towards Human-UAV Physical Interaction and Fully Actuated Aerial Vehicles

Dissertation

der Mathematisch-Naturwissenschaftlichen Fakultät

der Eberhard Karls Universität Tübingen

zur Erlangung des Grades eines

Doktors der Naturwissenschaften

(Dr. rer. nat.)

vorgelegt von

Sujit Rajappa

aus Nagercoil (Indien)

Tübingen

2017

Gedruckt mit Genehmigung der Mathematisch-Naturwissenschaftlichen Fakultät der Eberhard Karls Universität Tübingen.

Tag der mündlichen Qualifikation:	16.05.2017
Dekan:	Prof. Dr. Wolfgang Rosenstiel
1. Berichterstatter:	Prof. Dr. Heinrich Bülthoff, MPI, Tübingen
2. Berichterstatter:	Prof. Dr. Andreas Zell

To the best physicist I have known, ***Rajappa Johnrose***
and my loving mother, ***Rajani Rajappa****.*

Abstract

Unmanned Aerial Vehicles (UAVs) ability to reach places not accessible to humans or other robots and execute tasks makes them unique and is gaining a lot of research interest recently. Initially UAVs were used as surveying and data collection systems, but lately UAVs are also efficiently employed in aerial manipulation and interaction tasks. In recent times, UAV interaction with the environment has become a common scenario, where manipulators are mounted on top of such systems. Current applications has driven towards the direction of UAVs and humans coexisting and sharing the same workspace, leading to the emerging futuristic domain of *Human-UAV physical interaction*.

In this dissertation, initially we addressed the delicate problem of external wrench estimation (force/torque) in aerial vehicles through a generalized-momenta based residual approach. To our advantage, this approach is executable during flight without any additional sensors. Thereafter, we proposed a novel architecture allowing humans to physically interact with a UAV through the employment of sensor-ring structure and the developed external wrench estimator. The methodologies and algorithms to distinguish forces and torques derived by physical interaction with a human from the disturbance wrenches (due to e.g., wind) are defined through an optimization problem. Furthermore, an admittance-impedance control strategy is employed to act on them differently.

This new hardware/software architecture allows for the safe human-UAV physical interaction through exchange of forces. But at the same time, other limitations such as the inability to exchange torques due to the underactuation of quadrotors and the need for a robust controller become evident. In order to improve the robust performance of the UAV, we implemented an adaptive super twisting sliding mode controller that works efficiently against parameter uncertainties, unknown dynamics and external perturbations. Furthermore, we proposed and designed a novel fully actuated tilted propeller hexarotor UAV. We designed the exact feedback linearization controller and also optimized the tilt angles in order to minimize power consumption, thereby improving the flight time. This fully actuated hexarotor could reorient while hovering and perform 6DoF (Degrees of Freedom) trajectory tracking.

Finally we put together the external wrench observer, interaction techniques, hardware design, software framework, the robust controller and the different methodologies into the novel development of *Human-UAV physical interaction with fully actuated UAV*. As this framework allows humans and UAVs to exchange forces as well as torques, we believe it will become the next generation platform for the aerial manipulation and human physical interaction with UAVs.

Kurzfassung

Über die Interaktion zwischen Mensch und unbemannten Luftfahrzeugen, sowie vollständig steuerbar Luftfahrzeuge

Unmanned Aerial vehicle (UAV) sind in der letzten Zeit in den Fokus der Wissenschaft gerückt. Quadrotoren sind dabei eine Klasse von UAVs mit vier Propellern, die mehr und mehr für die Interaktion mit Objekten in der Luft verwenden werden. Neben Aufgaben zur Überwachung und zur Aufnahme von Daten, werden UAVs wegen ihrer Fähigkeit Orte zu erreichen und Aufgaben auszuführen, die für Menschen oder andere Roboter nicht möglich oder erreichbar wären, effizient eingesetzt. Nicht nur in Fällen in denen die Arbeitsbedingungen für Menschen ungünstig sind, sondern auch wenn es darum geht einen Arbeitsraum zu teilen und zusammenzuarbeiten, können UAVs mit auf diesen montierten Manipulatoren eingesetzt werden. Dies führt zu einem neuen Bereich der physikalischen Mensch-Maschine Interaktion.

In dieser Doktorarbeit beschäftigen wir uns zu Beginn mit der Schätzung von auf das UAV wirkenden, externen Kräften und Momenten mit Hilfe eines Störgrößenbeobachters. Dies wird durch einen auf 'generalized momenta' basierten Entwurf ermöglicht. Ein großer Vorteil dieses Entwurfes ist die Fähigkeit alle Berechnung in Echtzeit durchzuführen. Außerdem werden keine zusätzlichen Sensoren benötigt. Desweiteren präsentieren wir eine neueartigen Ring-Sensor zur physikalischen Mensch-Maschine Interaktion. Die Unterscheidung zwischen einer von außen wirkenden Kraft (wie z.B. Wind) die als Störung interpretiert wird und der gewollten Mensch-Maschine Interaktion wird als ein Optimierungs-Problem beschrieben. Die aus diesem Ansatz resultierenden Ergebnisse werden zusätzlich mit denen eines so genannten Admittanz-Impendanz Regler verglichen. Diese neuartige Architektur erlaubt eine sichere Mensch-Maschine Interaktion durch die Übertragung von Kräften. Gleichzeitig müssen einigen Gegebenheiten der Praxis Rechnung getragen werden. Dazu gehören die Berücksichtigung von Unsicherheiten, Störungen und nicht modellierter Systemdynamik. Aus diesem Grund werden zusätzlich Aspekte der Robustheit und die Unfähigkeit Momente aufzunehmen berücksichtig.

Um die Leistungsfähigkeit einer robusten Regelung des UAVs zu verbessern, verwenden wir einen adaptiven super twisting sliding mode controller, der eine Robustheit gegenüber Parameterunsicherheiten, unbekannter Systemdynamik und von außen wirkenden Kräften und Momenten verleiht. In einem weiteren Abschnitt dieser Arbeit präsentieren wir einen neuartigen Hexarotor mit sechs geneigten, auf den Energieverbauch optimierten Propellern. Für die Regelung dieses Prototyps implementierten wir einen exakten linearisierenden Regler mit Ausgangsrückführung. Ziel unserer Anstrengungen war es, eine vordefinierte Trajektorie folgen zu lassen und eine Drehung ohne eine Veränderung der Position zu ermöglichen.

Die Kombination aus Störgrößenbeobachter, Hardware Design, einer auf Robustheit und Leistung ausgelegten Reglerstrategie kompletiert diese Arbeit und bildet den Schluss. Durch die erbrachten Ergebnisse kann ein weiterer Schritt in die Richtung einer neuen Generation von UAVs zur physikalischen Mensch-Maschine Interaktion und die Basis für eine neue Generation von in der Luft agierenden Manipulatoren geschaffen werden.

Acknowledgments

This thesis work is the culmination of the many challenges and hardships in the last three years right from the start. It would not have been possible without the help of many good people who supported and accompanied in the last few years.

First and foremost, I express my sincere gratitude and respect to Prof. Dr. Heinrich H. Bülthoff for granting the great opportunity to work in the *Autonomous Robotics and Human Machine Systems* group and providing the valuable support throughout this thesis at MPI. This thesis work in aerial robotics would have turned unrealizable, if not for his timely intervention and motivation at various stages. Along similar lines, I would like to thank Prof. Dr. Andreas Zell for accepting me as his PhD Student within his group in University of Tübingen, for co-supervising and giving valuable advices.

I am very much grateful to my direct supervisor Dr. Paolo Stegagno for supervising the thesis, constant support at all time, continuous encouragement and valuable technical discussions. The last minute experiments and working nights during the deadlines will always be unforgettable memories to cherish. In the same manner, I also thank Dr. Antonio Franchi for his valuable inputs and supervision in the *Fully Actuated Hexarotor* project.

I thank my fellow colleagues from the fine mechanical and electronic workshop for their valuable help during the development of various parts for the UAV prototypes. I also thank the non-technical colleagues from the administration for always making my stay in MPI comfortable during this thesis period. Many thanks go to my group mates: Aamir, Burak, Christian, Caterina, Marcin, Matteo, Thomas, Marco, Saber, Jayjit and Eugen in the robotics group for the advices, discussions, get-togethers, fun and the good friendship. I also take this opportunity to thank my project collaborators Yuyi, Carlo and Markus for the valuable technical advices at various stages during this thesis.

My parents made a lot of sacrifices for me to pursue higher education and they were the source of constant motivation and encouragement to do a Ph.D. I especially thank my brother Dr. Rajit Rajappa who has been my much needed motivator and great adviser throughout this thesis period.

Last but not the least, I thank my wife Renee and daughter Iniya, for their constant support and big compromises they made while I stayed long hours away at work. It would not have been possible without their encouragement.

Tübingen, January 2017
Sujit Rajappa

Contents

1 Introduction **1**
1.1 Aerial Robotics for Civilian Purpose 1
1.2 UAV Platform 3
1.3 Motivation 6
1.4 Characteristics and Challenges 8
1.5 Objectives and Outline of the Thesis 9

2 External Wrench Estimation **13**
2.1 Introduction 13
2.1.1 Related Works 14
2.1.2 Methodologies 15
2.2 Preliminary System Descriptions 16
2.2.1 Model of the External Wrench 18
2.3 External Wrench (Force/Torque) Observer 19
2.4 Disturbance Compensator in Near-Hovering Control 22
2.4.1 Standard near-hovering control 22
2.4.2 Calculation of roll (ϕ_c) and pitch (θ_c) compensation 24
2.5 Simulations and Analysis 26
2.5.1 Hovering with constant wind disturbance 26
2.5.2 Hovering with varying wind disturbance 29
2.5.3 Trajectory Tracking with constant wind disturbance 29
2.5.4 Trajectory Tracking with varying wind disturbance 32
2.6 Experimental Validation 33
2.6.1 Force Estimation Experiment 34
2.6.2 Torque Estimation Experiment 34
2.7 Discussions and Possible Extensions 37

3 Novel Architecture for Human-UAV Physical Interaction **39**
3.1 Introduction 39
3.1.1 Related works 40
3.1.2 Methodologies 42
3.2 Problem Setting 44
3.2.1 Preliminary System Descriptions 44
3.2.2 Extended Model of the External Wrench 44
3.2.3 Force/Torque detectors 45

3.3 System Architecture 46
3.4 Hardware-Software Design 46
3.5 Estimation of the External, Interactive and Disturbance Wrenches . . . 49
3.5.1 Estimation of the External Wrench 49
3.5.2 Estimation of the disturbance and interaction wrenches 50
3.6 Control 53
3.6.1 Admittance Control 53
3.6.2 Trajectory Tracking Control with Wrench Feedforward 55
3.7 Hardware-in-the-loop Physical Simulations 57
3.8 Experimental Validation 60
3.8.1 Continuously Pushing, Sudden Impact and Multiple PoCs . . . 61
3.8.2 Trajectory tracking during human interaction with varying stiffness 63
3.9 Discussions and Possible Extensions 64

4 Robust Adaptive Super Twisting Control 67
4.1 Introduction 67
4.1.1 Related Works 68
4.1.2 Methodologies 68
4.2 Preliminary System Descriptions 69
4.2.1 Dynamic System Model 69
4.2.2 Regular Control Form 71
4.2.3 Uncertainties 74
4.3 Control 75
4.3.1 Adaptive Super Twisting Control 76
4.3.2 Feedforward Control 78
4.4 Physical Simulations 78
4.4.1 Experimental Setup 79
4.4.2 Robustness of ASTC 81
4.4.3 Comparison of ASTC and STC 81
4.5 Discussions and Possible Extensions 83

5 Fully-actuated Hexarotor Aerial Vehicle with Tilted Propellers 85
5.1 Introduction 86
5.1.1 Related Works 86
5.1.2 Methodologies 87
5.2 Design and Modeling 88
5.2.1 Static System Description 88
5.2.2 Equations of Motion 91
5.3 Control Design 93
5.3.1 Exact Feedback Linearization and Decoupling Control 94
5.4 A Preliminary Prototype 96
5.4.1 Discussion on the Invertibility of $J(\alpha_h, \beta_h)$ 97

5.4.2 Optimization of α_h and β_h . . . 98
5.5 Simulations and Analysis . . . 101
5.5.1 Reorienting while hovering with external disturbance . . . 101
5.5.2 6 DoF trajectory tracking . . . 103
5.6 Hexarotor Prototype . . . 103
5.6.1 Hardware . . . 106
5.6.2 Software . . . 107
5.7 Experimental Validation . . . 107
5.7.1 Hovering and Reorienting . . . 108
5.7.2 6 DoF Trajectory Tracking . . . 110
5.8 Discussions and Possible Extensions . . . 112

6 Human-UAV Physical Interaction with a Fully Actuated UAV **115**
6.1 Introduction . . . 116
6.1.1 Related Works . . . 118
6.1.2 Methodologies . . . 118
6.2 Design and Modeling . . . 118
6.3 UAV-HRPI System . . . 120
6.3.1 Interaction Wrench Observer . . . 121
6.3.2 Admittance Control . . . 122
6.3.3 Adaptive Super Twisting Control . . . 123
6.4 Simulations and Analysis . . . 129
6.4.1 Human-Robot Physical Interaction with a Fully Actuated UAV . 129
6.4.2 Comparison of Underactuated and Fully-Actuated dynamics . . 134
6.5 Discussions and Possible Extensions . . . 134

7 Conclusions **137**
7.1 Future Research Directions . . . 139

A Technical Computations **141**
A.1 Computation of Fully Actuated Hexarotor Model . . . 141
A.1.1 Translational Dynamics . . . 142
A.1.2 Rotational Dynamics . . . 145

B Mechanical Schematics **151**

Nomenclature **157**

Symbols **159**

Abbreviations **161**

Bibliography **163**

Chapter 1

Introduction

1.1 Aerial Robotics for Civilian Purpose

Aerial robotics is a branch of robotics whose primary object of study are Unmanned Aerial Vehicles (UAV). Fairly recently, UAV based research has been within the closed hands of governments being used mainly for military purposes. In the last two decades, academic researchers (mainly roboticists) started working with flying robots and thus opening the gates of UAVs for civilian applications. UAVs for civilian mission are making lot of ground. This can be seen with the growing number of conferences, workshops, forums, publications and start-ups in aerial robotics. Industrial investments has increased exponentially for the employment of aerial vehicles for industrial production, surveillance, maintenance, manipulation, etc., which were previously performed only by manipulators and ground robots[1]. In line with the demanding innovation and necessity for using the aerial robots, the European Commission has invested tremendously during the last decade in many of completed and ongoing projects, namely: AIRobots[2], ARCAS[3], Eurathlon[4], Hephestos[5], sFly[6], Valeri[7], Sherpa[8], etc., reiterating again to the tax-paying public about the direction of future technology.

According to the survey from the Center for Research on Globalization[9], apart from the use in military applications, UAVs are recently getting utilized a lot for a wide range of civilian applications. Particularly, Micro Aerial Vehicle (MAV) applications include disaster response, commercial delivery, exploration, archaeological surveying, environmental study (climate study, storm monitoring, mapping glaciers), security (public safety, surveillance, crowd monitoring), law enforcement (traffic management, search and rescue operation, aiding hostage situation), firefighting, health care (medical emergency and

[1] http://www.airborne-robotics.com/en
[2] http://cordis.europa.eu/project/rcn/93629_en.html
[3] http://www.arcas-project.eu/
[4] http://cordis.europa.eu/project/rcn/106965_en.html
[5] http://www.hephestosproject.eu/confluence/dashboard.action
[6] http://www.sfly.org/
[7] http://www.valeri-project.eu/
[8] http://www.sherpa-project.eu/sherpa/
[9] http://www.globalresearch.ca/unmanned-aerial-vehicles-uav-drones-for-military-and-civilian-use

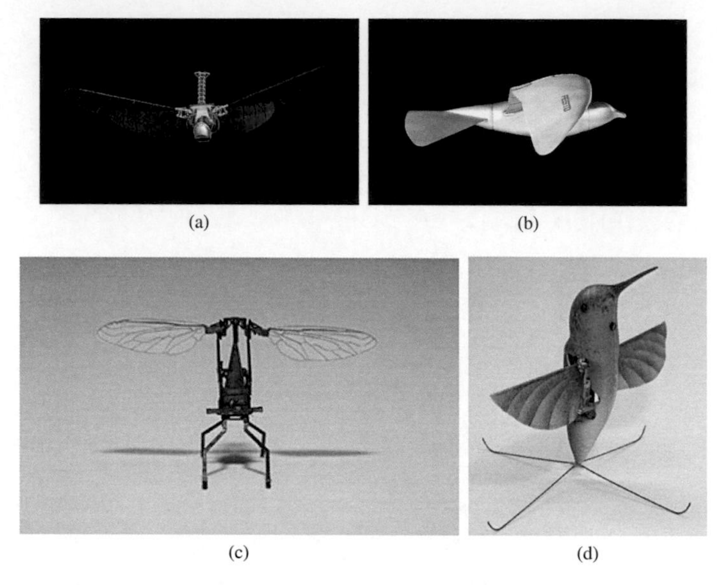

Figure 1.1: Example of Flapping wing UAVs. 1.1(a): BionicOpter. 1.1(b): Festo Smart Bird. 1.1(c): Harvard RoboBee. 1.1(d): Nano Hummingbird. *Source: www.google.com.*

delivery), farm management (spraying, watering, crop management), aerial photography, entertainment (film shooting), thermography survey, land inspection, pipeline inspection, marine life monitoring, wildlife conservation, etc.

Previously, research on flying robots was mainly focused on modeling (Erginer and Altug, 2007; Bresciani, 2008), system identification, state estimation (Mokhtari and Benallegue, 2004; Abeywardena *et al.*, 2013), trajectory planning (Bouktir *et al.*, 2008) and control (Mistler *et al.*, 2001; Bouabdallah and Siegwart, 2007). This paved the way for the development of novel UAV designs, design and employment of various state estimation techniques and control methodologies even though the application intend was poorly defined. They were used for a variety of purposes mostly in the direction of inspection and surveillance. At a time when new advancements was envisioned for aerial robots, it came from the sensor technology. Computer vision in UAVs became a domain within aerial robotics with still many open problems and investigated by many research groups.

The application direction became wider and moved towards UAV manipulation whose preciseness was influenced by vision based navigation and control. Visual servoing (Guenard *et al.*, 2008; Mebarki *et al.*, 2015) became a popular direction and topic for major investigation in UAVs. This led to manipulators being installed on top of aerial vehicles for various manipulation tasks (Mebarki *et al.*, 2014; Montufar *et al.*, 2014; Suarez *et al.*, 2015; Kim *et al.*, 2016). Aerial manipulation is a difficult mission in itself because the force exchange could affect the stability of the UAV. Many new techniques are still being developed for the accomplishment of this task. Furthermore, this also steered the applications from UAV manipulation towards interaction with the environment (Fumagalli *et al.*, 2014; Darivianakis *et al.*, 2014; Yüksel *et al.*, 2015). UAV interaction opened up unforeseen possibilities in civilian applications using aerial vehicles for tasks which were previously considered infeasible, since UAVs have the advantage of reaching places which were not accessible for humans or ground robots. Hence, physical contact with the environment started gaining importance in the scientific community making *UAV manipulation and interaction with environment* one of the hottest topics within UAV enthusiasts.

"Human-UAV Physical Interaction", exchange forces between UAVs and human operators, is the next big futuristic step which we take for employing UAVs in civilian applications. It is based on the idea of humans and UAVs coexisting in the same workspace to accomplish endeavors. In such a scenario, physical contact is unavoidable and necessary. A detailed study on this is discussed through in this dissertation.

1.2 UAV Platform

Before introducing the UAV platform that is used for various civilian based application, in this section we highlight the many categories under which UAVs could be divided. UAVs are mainly classified based on the shape and size, working principle or applications. The differentiation based on size helps to group them under heavy (mostly used in military applications) and light aerial vehicles. Though it is not necessary that only light aerial vehicles be used for civilian application, this is the current trend based on safety reasons. Micro Aerial Vehicles (MAV), small size UAVs used by almost all the civilian application oriented research groups, also fall under this category

A second classification criterion is to distinguish the robots based on the principle of flight mechanism. Most of the civilian purpose UAVs designed by roboticists fall under one of the following categories: (i) fixed wings, (ii) rotary wings, and (iii) flapping wings. *Bio-inspired aerial robot* models mostly fall under the flapping wings working principle where the flight dynamics are mostly defined like as in a bird or insect. Most of these novel models try to imitate the complex flight mechanism which nature has obtained through millions of years of evolution. Flapping wing based UAVs are an upcoming direction and are mostly in the early research stage. Some examples of flapping wing UAV models (Festo Smartbird, NanoHummingbird, BionicOpter, etc.) can be seen in Fig. 1.1.

Fixed wing UAVs are well known because their flight principle is similar to commercial airplanes. Bigger size fixed wing UAVs require proper take-off/landing strip and are used mostly by military for surveillance, reconnaissance, etc. Some examples such

Figure 1.2: Fixed wing UAV examples. 1.2(a): Neuron. 1.2(b): Hermes. 1.2(c): Predator. 1.2(d): Reaper. 1.2(e): Tianying. 1.2(f): X8. 1.2(g): Arcturus T-20. 1.2(h): GLG8. *Source: google.com.*

as Predator, Reaper, etc., are shown in Fig. 1.2(a-d). The advantages of using such vehicles usually are that they can operate at very high speed, reach high altitude, travel long distance and be airborne for longer flight time because they are usually powered by combustible hydrocarbons. Small size fixed wing UAVs are sometimes used for civilian application such as photography, field surveillance, etc., and could be vehicle or hand launched. Some models of such vehicles such as X8 are shown in Fig. 1.2(e-h).

Lastly, rotary propeller UAVs working principle is similar to helicopters, where there are propellers blades to create the thrust required for being airborne. The number of needed propellers differs and depends upon the degrees of freedom (DoF) of the vehicle. Within rotary wing UAVs, they could be divided as well based on their size. The heavy size ones come in different models for military grade such as, for example: Firescout, Skeldar, etc., as seen in Fig. 1.3. The small size civilian application ones are mostly with four propellers and light weight such as, e.g., DJI Phantom, Asctec Hummingbird, etc., as also seen in Fig. 1.3(d). The biggest advantages of using rotary propeller UAVs are their vertical take-off and landing (VTOL) capability, hovering capability, easily deployable, high maneuverability, fly indoor and mostly importantly could be easily setup for research applications.

Quadrotor UAVs used here comes under the rotary wing, VTOL vehicle and small size MAV used mostly for civilian application. They have four propellers that are spaced at 90° apart and are parallel to one another in the same plane such that the forces are always generated along the Z-axis of the vehicle. The quadrotor works by simple mechanics unlike the helicopter which needs to change the angle of the rotor blades for stable closed-loop control flight. Here, each of the fixed rotor generates the required force and torque. Therefore, it can generate roll, pitch, yaw torque and thrust by varying the rotational angular velocity of the individual propellers. This also requires therefore precise control of the propeller speed through the motors for stable hovering flight.

Since quadrotors come under the rotary wing class, they don't require much space and they can be designed in different compact sizes. Moreover the VTOL capability allows them to be deployed easily in any environment without much hurdle. These quadrotors normally have a mass of around 1 kg and they are battery powered allowing a flight time of $\simeq 15\ min$.

The propellers of the quadrotors are fixed such that alternate propellers are rotating in opposite direction, i.e. clockwise and anti-clockwise direction. The propeller rotation generates drag torques. These torques could be internally balanced and stabilized if equal and opposite torques are generated by even number of propellers. This is the reason for the clockwise and anti-clockwise arrangement of the propellers even though the forces are always generated in the same direction.

As mentioned earlier, quadrotors can generate three torques (roll, pitch, yaw) in all directions and thrust by means of the control inputs to its four propellers. Therefore, with only four degrees of freedoms, quadrotors are underactuated systems in which position in space and yaw (rotation around the vertical axis) can be freely achieved, but the attitude (i.e., desired roll or pitch) cannot. The MK-quadro from the Mikrocopter which has been

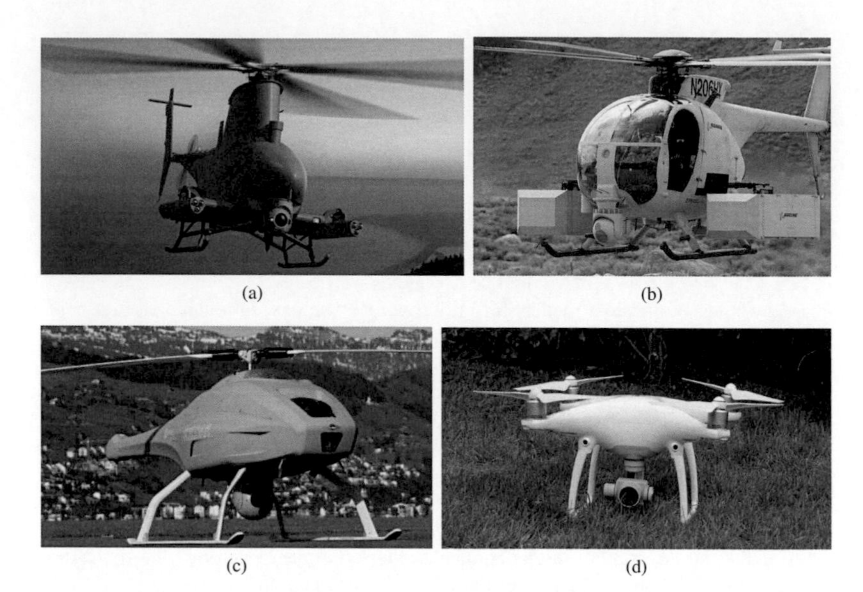

Figure 1.3: Example of Rotary wings UAVs. 1.3(a): Fire Scout. 1.3(b): Little Bird. 1.3(c): Skeldar. 1.3(d): Phantom. *Source: www.google.com.*

used in thie development of this work is shown in Fig. 1.4. The dynamic model along with frame reference of the quadrotor is introduced later in Sec. 2.2 and the hardware setup is detailed in Sec. 2.6.

1.3 Motivation

Summarizing from Sec. 1.1, there are several motivations for considering working with UAVs and physical interaction:

- Right from the start, UAVs have always been fascinating. Even though the general public have mostly stayed away from UAV related research for many years, because of its known use in military related application, there is no doubt about the capability of UAVs. They are capable of executing tasks autonomously with speed and precision in environments that are not reachable and considered dangerous for humans to operate in.
- Most of current UAV applications involve flying in human populated areas where interaction with the robot could be a common scenario and possibility. The hottest

Figure 1.4: MK-Quadro from Mikrokopter.

research topic currently in aerial robotics is UAV manipulation and interaction with the environment. As the application domain is getting wider, the current trend is moving towards physical human interaction with UAVs for task accomplishment. Some form of interaction between humans and UAVs will become necessary in order to transfer information through physical contact and force exchange.

- Cooperation between humans and UAVs has always been envisioned. *Assistive Robotics* has in itself become a research direction within robotics because humanity has always benefited from robots. Together humans and UAVs can successfully and effectively accomplish a wide range of task which have been proved through the many applications with manipulators and ground mobile robots. The civilian purpose UAVs have the ability to coexist with humans if technology addresses the safety issue.

- Safety must be given the primary importance. When it comes to UAVs, this needs to be addressed from multiple directions. Firstly, the technology must be autonomous and fool proof enough for safe operation. UAVs must possess more interactive based safe behavior states similar to the hovering state. Moreover, the mechanical hardware setup of the UAV needs to be better designed to allow human-UAV interaction. Current typical UAV platforms have underactuated dynamics, which is in the first place not ideal for interaction tasks because of the inability to torque exchange and define a desired orientation in free space.

1.4 Characteristics and Challenges

Considering the application domains and the research direction mentioned in Sec. 1.1 and its motivations and drawbacks explained in Sec. 1.3, physical interaction between UAVs and humans can give raise to different set of challenges and problems, such as safety, stability of the UAV, ethical and social implications, control, etc.

The most important challenge for researchers working with the aerial robotics in general and UAVs in particular is the inability to convince the general public that such systems could be friendly and be used for civilian applications as well. The discussions related UAVs have always been hostile outside the research domain. While researchers are not the primary reason for such a perception, it would take lot of efforts from all stakeholders and the civil application exposure over a period of time to change this pugnacious mindset.

The next biggest challenge is the lack of well defined rules and regulations for UAV flights in human populated areas for civilian application. Though certain governments and regulators have started to take notice of this, in general there is currently no commonly regulated system in place. This have caused a lot of setbacks for UAV based research in using to the fullest the currently available technological advancements for the betterment of the society.

Technologically, there are many open problems that are still to be solved. The flight time with the current UAV platform has been preventing efficient UAV employment in many ways. The average flight time for a typical quadrotor platform is between $10-20$ *mins* and this varies inversely as the payload capacity is increased. Battery technology has been widely spoke about and has gained a lot of research attention in recent years from a wide range of application industry. The safety of employing UAVs in human populated regions is still debatable. None of the commercially available UAV providers have been able to guarantee that these systems are safely handled by an user except for experienced pilots. The ability of unmanned vehicles to automatically adhere to safe operation is still under investigation both in the automobile and aerial robotics domain.

Now with the application and research interests moving towards UAV interaction with the environment, new questions are revealed. Technologies and concepts which were previously not investigated in aerial robotics but were relevantly studied in other domains of robotics are revisited and adapted for their application with UAVs. Since the motivation and the topic of this thesis aligns with UAV interaction with the environment, it makes sense to highlight some of the relevant challenges related to it, namely:

- The development of external wrench (i.e., external forces and torques acting on the robot) estimators for aerial vehicles was not deemed important though considered useful. Now with force and torque exchanges being a primary research objective, there is a dire need for onboard wrench observers that could work during flight.
- Linear controller on UAVs have always been more efficient than non-linear controllers due to fine-tuning and operation in known structured environment. Robust

controllers for aerial vehicles are now required, since in real applications external perturbations are quiet varied, not systematic and unpredictable along with the unknown and uncertainty in the workspace environment.

- Sensor integration for real-time aerial perception of the object and the environment is now relevant. Computer vision has accomplished high levels of accuracy in manipulators. In aerial vehicles, with limited computational power and stringent real-time requirements, the biggest challenge is that the most of these techniques are computationally very heavy. Therefore, sensor integration and related techniques are still an open issue.

- The philosophy of human physical interaction with robots. This has proved beneficial in manipulators for the accomplishment of various tasks. However in aerial robots this idea is yet to take shape because of the lack of an interaction methodology and framework.

- The well known problem of underactuation in quadrotor aerial vehicles which has drastically limited the ability of aerial interaction and manipulation using UAVs. This has not played any role in spoiling sport problems as trajectory tracking and surveillance. But when it comes to interaction related application tasks, this becomes a serious issue that hinders the capabilities and overall stability of the aerial vehicle.

1.5 Objectives and Outline of the Thesis

The discussion on the characteristics, of the human-UAV physical interaction and the fully actuated UAVs in Sec. 1.4, gives a brief glimpse on the importance of this topic along with its existing several challenges. The goal of this Ph.D. thesis is mainly to attempt solving few of the basic specific problems that hinder the initial steps towards the futuristic topic of UAV-HRPI (Human Robot Physical Interaction). Moreover solving theses problems gives an insight on the bigger possibilities that are open from the point of view of technology as well as civilian application with UAVs. Summarizing, important contributions of this thesis are:

- ***External Wrench Estimation:*** The problem taken into account is how to produce an estimate of the external wrench acting on a UAV using only the existing system states and measurement without adding additional sensors. This is important for the hardware platform of a civilian purpose quadrotor UAV where the payload capacity and computational power directly affects the flight time. Moreover, such an observer should have a robust convergence property since it directly could affect the UAV's stable hovering state.

- ***Human-UAV Physical Interaction:*** In recent years, Human Robot Interaction (HRI) have been very much talked about. HRI with UAVs has also been studied in the last decade. But Human Robot Physical Interaction (HRPI) with UAV is futuristic. The underlying problem is that there is no hardware setup or software framework to attain UAV-HRPI. One of our objective is therefore to develop such a technology so that humans can physical interact through forces and torques safely with UAVs.

- ***Robust Control:*** Linear control techniques are so popularly used for UAVs because of their simplicity and familiar tuning. But UAV applications are moving from well defined indoor setup towards unknown outdoor environment prone to uncertainties and perturbations. The solution of providing a robust non-linear controller for UAVs which can be utilized in all application scenarios is one of the objectives taken in hand.

- ***Full-Actuation:*** Underactuation has always stayed as an issue with the most commonly used quadrotor UAV platform. The research solutions provided always had the controllability problem making it unfeasible for aerial interaction and manipulation. Moreover, UAV-HRPI with underactuated systems limits the interaction to only force exchanges, but not torques. One of the objectives of this thesis is to develop a fully actuated UAV that, having simple control properties, can be used as a next generation UAV platform for aerial manipulation as well as physical interaction.

The rest of the thesis is organized as follows:

- In Chap. 2 it is introduced the preliminaries of the UAV system dynamics together with the methodology for the external wrench estimation and the design of a disturbance compensator factor that needs to injected in the controller for stabilization of the UAV while subjected to external forces/torques.

- In Chap. 3 it is designed a novel hardware architecture for Human-UAV physical interaction along with the methodology for the separation of interaction wrenches due to contact from aerodynamic disturbances. Then it is developed a control framework which allows humans to provide intuitive force commands to the UAV.

- In Chap. 4 it is implemented a robust adaptive super twisting sliding mode controller for UAV which posses the properties of compensating for uncertainties in system dynamics and unknown perturbations using its non-linear control action.

- In Chap. 5 it is proposed and developed a novel hexarotor prototype with tilted propellers which has the capability to act as a fully-actuated UAV with 6 Degrees of Freedom (DoF) that can generate force as well as torque in any arbitrary direction.

- In Chap. 6 it is developed an adaptive super twisting controller and interaction wrench estimation methodology that is implemented in the 6 DoF fully-actuated hexarotor with tilted propellers. It is also adapted to the new system dynamics the admittance control framework developed in Chap. 3 for humans to intuitively exchange interaction forces and torques.

- Chapter 7 concludes this thesis with a summary of the contributions and a discussion on the implication of human-UAV physical interaction using a fully actuated aerial vehicle.

Chapter 2

External Wrench Estimation

With the application direction moving towards UAV physical interaction with the environment and manipulation, the knowledge of the external wrench (force/torque) acting on a UAV becomes a necessity to design and implement stable and efficient controllers allowing the UAV to exert a predetermined force. The easiest way would be to integrate a force/torque sensor. This method has been efficiently used in many ground robots/manipulators, but on UAVs this is not so advantageous. Unlike ground robots, additional weight will directly affect the aerial vehicles performance along with stability apart from the cost, power consumption and flight time concerns.

What could be the best way to obtain wrench information? Can it be easily implemented in an aerial robot without increasing the computation time, power consumption and payload capacity? Can it be effectively used for real-time UAV application?

In this chapter we answer these questions by implementing a methodology of external wrench observer using a *generalized momenta residual-based* technique which can be efficiently implemented on a UAV.

The discussion presented in this chapter is based upon the work that I have done under the supervision of Dr. Paolo Stegagno during stage-I European Robotics Challenges (EUROC)[1] and is partly to appear in Rajappa *et al.* (2017a).

2.1 Introduction

Because of their privileged point of view and their ability to work on difficult terrains, aerial robots can be employed for inspection and maintenance of otherwise hardly accessible areas both in indoor and outdoor scenarios. *Aerial manipulation and UAV interaction with the environment* represent one of the most important topics of aerial robotics research, with tasks ranging from simple inspection to complex interaction such as mobile manipulation (see e.g., Orsag *et al.* (2013); Lippiello and Ruggiero (2012)), aerial grasping (see e.g., Pounds *et al.* (2011); Lindsey *et al.* (2011)) or exert forces on objects (Gioioso *et al.* (2014b)).

[1] http://www.euroc-project.eu/

In most cases, such applications often bring the UAVs to extreme situations, in which either external disturbances as wind or the required task itself as physical interaction with the environment cause external forces and torques acting on them. The quadrotor, the typical UAV research and commercial platform, have only 4 control inputs compared to the 6 degrees of freedom (DoF) required to pose (position and orient) the UAV in free space and therefore come under the class of under-actuated robots. Furthermore, the translational and rotational system dynamics of quadrotors are coupled. All these reasons point that quadrotors are more vulnerable to external wrenches (model uncertainty, external perturbations or interaction wrench) than any other class of robots. Sometimes during interaction applications that are using mechanical manipulator designs (Yüksel *et al.*, 2014; Gioioso *et al.*, 2014b), this vulnerability is ignored. Nevertheless, in most cases applications are carried out relying on the control performances for the stability during interaction.

In order to overcome this shortcomings, it becomes very important and necessary to employ different approaches. The typical control strategy involves devising a robust controller (such as feedback linearization (Voos, 2009), backstepping (Madani and Benallegue, 2006), sliding mode (Xu and Ozguner, 2006), etc.), which guarantees stability and asymptotic trajectory tracking, to counteract the generated system state errors. However, when it comes to intuitive interaction or exchange of wrenches, it becomes necessary to have the knowledge of the external wrench. With the estimate of the wrench and depending on the application along with the source of the external wrench, the control strategy can also be applied efficiently.

In this chapter it will be introduced the general UAV system dynamics and then discussed the methodology to observe the external force/torque wrench acting on the aerial vehicle. This observer in real-time does not require any additional sensor, mechanical designs or payload but works by exploiting the already available Inertial Measurement Unit (IMU) data, UAV position and velocity estimates and control inputs generated by the controller.

2.1.1 Related Works

Historically, the utilization of an observer to estimate the unknown system states has been a common practice in control theory and this is understood from the available literature with the wide range of different methodologies (Luenberger (1979); Khalil (2002)). In practice, the employment of this observer varies differently depending on the application and the available system states. The implementation in robotics has been successful and has been commonly used in ground robots (Erlic and Lu (1993); Chen *et al.* (2000); Luca *et al.* (2007)). However, the execution of an observer in aerial robotics and particularly quadrotor UAV has been always with various degrees of success because of the limited on-board payload and computation capacity. Here the idea is to use an observer to estimate the unknown external forces and torques that are acting on the UAV.

The estimation of the environmental forces acting on the rigid body manipulator during contact tasks was observed by Hacksel and Salcudean (1994). The measured position, orientation and the actuation forces were utilized to observe the rigid body velocities for control. In UAVs, Bellens *et al.* (2012) investigated the external wrench estimation for flying robots in the context of hybrid pose/wrench control. However this method was an offline measurement of the forces and torques generated by the UAV. Nguyen and Lee (2013) used directly a force sensor to estimate the external wrench and utilize it for a tool operation with a quadrotor. Current research is proving more useful the employment of an observer to estimate the external force/torque instead of mounting a force/torque sensor or relying on external sensors. This is because the additional onboard sensor would increase the payload indirectly compromising the flight time of the quadrotor, while external sensors would limit the field of application to structured environments.

From the control point of view different approaches were suggested, mainly without estimation, to counteract the external wrench considering them as disturbances. Albers *et al.* (2010) used a force control approach with an external feedforward signal to decrease the influence of disturbance. Among the many control approaches, the adaptive methods were more robust, because the controller has the ability to adapt depending upon the magnitude of the external wrench (see e.g., Roberts and Tayebi (2011); Palunko *et al.* (2012); Antonelli *et al.* (2013)).

Recently, Augugliaro and D'Andrea (2013) proposed a wrench estimation method based on an unscented kalman filter for the linearized model of a quadrotor UAV. An alternative Lyapunov-based nonlinear observer for estimating the external forces applied has been proposed by Yüksel *et al.* (2014) and numerically validated. However, this approach was not robust enough to precisely track a high frequency rapidly varying wrench. A residual based approach had been used in Collision/Fault Detection and Identification (FDI) methodology (Takakura *et al.*, 1989) for robotic manipulator arms for the safe operation. Further it was improved based on the generalized momenta of the robot (De Luca and Mattone, 2003) and used in manipulator arms. Our method is inspired from the FDI technique for manipulators to be used for UAVs. Recently, a modified version of this method was utilized (Ruggiero *et al.*, 2014; Tomic and Haddadin, 2015) for different quadrotor UAV application at the same time as our development.

2.1.2 Methodologies

Therefore to summarize the work presented in this chapter:

1. it is presented a methodology without the utilization of onboard or external sensors to observe the force/torque external wrench. This momenta based residual FDI technique, used earlier for collision detection, is exploited and modified to be efficiently used in aerial robots for 6 Dimensional (6D) wrench estimation;

2. it is computed the feedforward compensation factor utilizing the estimated wrench to reject the arising disturbance in the system dynamics that affects the UAV's stability;

3. it is discussed the implementation of the compensation factor in the existing control architecture for UAVs and validated.

2.2 Preliminary System Descriptions

The quadrotor UAV is modeled as a rigid body moving in 3D space as mentioned earlier in Sec. 1.2. In order to estimate the external disturbance wrench (force/torque) that are acting on the UAV, it is important to define the quadrotor system dynamics in terms that can be used for the disturbance observer. The world inertial frame in which the quadrotor flies is denoted as $\mathcal{F}_W : \{\boldsymbol{O}_W, \vec{\boldsymbol{X}}_W, \vec{\boldsymbol{Y}}_W, \vec{\boldsymbol{Z}}_W\}$ and the body frame attached to the quadrotor is defined as $\mathcal{F}_B : \{\boldsymbol{O}_B, \vec{\boldsymbol{X}}_B, \vec{\boldsymbol{Y}}_B, \vec{\boldsymbol{Z}}_B\}$, where $\boldsymbol{O}_\star$ is the center of the frame $\mathcal{F}_\star$ and $\vec{\boldsymbol{X}}_\star, \vec{\boldsymbol{Y}}_\star, \vec{\boldsymbol{Z}}_\star$ are the three principal axes of frame $\mathcal{F}_\star$. Here $\boldsymbol{O}_B$ coincides with the quadrotor Center of Mass (CoM). The visualization of the different frames can be seen in Fig. 2.1.

Let $\boldsymbol{p}_W = [x \ y \ z]^T \in \mathbb{R}^3$ describe the position of $\boldsymbol{O}_B$ in $\mathcal{F}_W$ and let $\boldsymbol{\Theta}_W = [\phi \ \theta \ \psi]^T \in \mathbb{R}^3$ be the standard roll, pitch and yaw angles respectively which describe the orientation of $\mathcal{F}_B$ in $\mathcal{F}_W$, with $\phi, \theta \in [-\pi/2, \pi/2]$ and $\psi \in [0, 2\pi]$. The basic quadrotor states are therefore

$$\boldsymbol{\xi}_W = \left[\boldsymbol{p}_W{}^T \quad \boldsymbol{\Theta}_W^T\right]^T = \left[x \quad y \quad z \quad \phi \quad \theta \quad \psi\right]^T. \tag{2.1}$$

Let $\boldsymbol{R}_B^W = \boldsymbol{R}_z(\psi)\boldsymbol{R}_y(\theta)\boldsymbol{R}_x(\phi) \in \mathbb{R}^{3\times 3}$ represent the rotation between $\mathcal{F}_B$ and $\mathcal{F}_W$:

$$\boldsymbol{R}_B^W = \begin{pmatrix} c_\psi c_\theta & c_\psi s_\theta s_\phi - s_\psi c_\phi & c_\psi s_\theta c_\phi + s_\psi s_\phi \\ s_\psi c_\theta & s_\psi s_\theta s_\phi + c_\psi c_\phi & s_\psi s_\theta c_\phi - c_\psi s_\phi \\ -s_\theta & c_\theta s_\phi & c_\theta c_\phi \end{pmatrix} \tag{2.2}$$

where $c_\star = \cos(\star)$, $s_\star = \sin(\star)$ and $\boldsymbol{R}_z$, $\boldsymbol{R}_y$, $\boldsymbol{R}_x$ denote the 3×3 fundamental rotation matrices around the Z, Y and X axes respectively. In order to benefit from the utilization of minimal system states (Stegagno *et al.*, 2013), the horizontal frame is defined as $\mathcal{F}_H : \{\boldsymbol{O}_H, \vec{\boldsymbol{X}}_H, \vec{\boldsymbol{Y}}_H, \vec{\boldsymbol{Z}}_H\}$ such that $\boldsymbol{O}_H \equiv \boldsymbol{O}_B$, $\vec{\boldsymbol{Z}}_H \parallel \vec{\boldsymbol{Z}}_W$ and $\psi_H = 0$, where ψ_H is the yaw angle of the UAV expressed in $\mathcal{F}_H$. Then, the rotation matrix between $\mathcal{F}_W$ and $\mathcal{F}_H$ is $\boldsymbol{R}_H^W = \boldsymbol{R}_z(\psi)$ and the rotation matrix between $\mathcal{F}_H$ and $\mathcal{F}_B$ is $\boldsymbol{R}_B^H = \boldsymbol{R}_y(\theta)\boldsymbol{R}_x(\phi)$. Hence the state of the UAV in $\mathcal{F}_H$ is

$$\boldsymbol{\xi}_H = \left[\boldsymbol{p}_H{}^T \quad \boldsymbol{\Theta}_H^T\right]^T = \left[0 \quad 0 \quad 0 \quad \phi \quad \theta \quad 0\right]^T. \tag{2.3}$$

The actuation system of the quadrotor consists of four motor-propeller pairs attached to four rigid arms. This configuration allows to command independently three torques $\boldsymbol{\tau} = [\tau_x \ \tau_y \ \tau_z]^T \in \mathbb{R}^3$ around the three axes $\vec{\boldsymbol{X}}_B$, $\vec{\boldsymbol{Y}}_B$, $\vec{\boldsymbol{Z}}_B$ and one force $\rho \in \mathbb{R}_0^+$ called thrust along $\vec{\boldsymbol{Z}}_B$. From the control point of view recent research proved convenient (Lee *et al.*, 2010) to define the translational dynamics of quadrotors in terms of the world frame $\mathcal{F}_W$

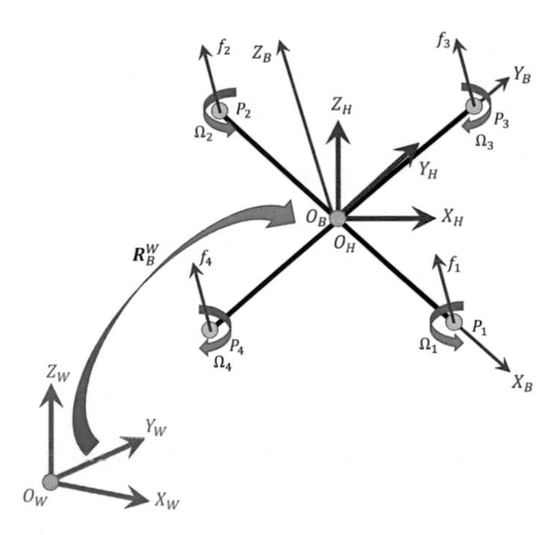

Figure 2.1: Schematic figure of the different frame reference. For the ith propeller: P_i is the propeller setup, Ω_i is the spinning velocity along with its direction (light blue), f_i is the force generated. $\mathcal{F}_W$ (green) is the world inertial frame, $\mathcal{F}_B$ (blue) is the quadrotor body frame and $\mathcal{F}_H$ (red) is the horizontal frame.

and the rotational dynamics in terms of the quadrotor body frame $\mathcal{F}_B$. Therefore the generalized velocity vector states are expressed as:

$$\boldsymbol{\zeta} = \left[\dot{\boldsymbol{p}}_H^T \quad \boldsymbol{\omega}_B^T\right]^T, \tag{2.4}$$

where $\dot{\boldsymbol{p}}_H \in \mathbb{R}^3$ is the linear velocity of the quadrotor in $\mathcal{F}_W$ expressed in $\mathcal{F}_H$ and $\boldsymbol{\omega}_B = [p \;\; q \;\; r]^T \in \mathbb{R}^3$ is the angular velocity of the quadrotor in $\mathcal{F}_W$ expressed in the $\mathcal{F}_B$.

To reduce the complexity of the arising quadrotor model, we consider the following standard assumptions:

Assumption 2.1:

$\mathcal{F}_B$ is aligned with the principal axes of the quadrotor.

Assumption 2.1 ensures that the inertial matrix $\boldsymbol{I}_B$ is diagonal.

Assumption 2.2:

The inertial and gyroscopic effects arising from propellers and the motors are rejected by the feedback nature of the controller considering them as second-order disturbances.

With the above mentioned assumptions and using the standard Newton-Euler equations of motion, the dynamical model of the quadrotor can be written as (Lee *et al.*, 2013)

$$m\ddot{\boldsymbol{p}}_W = -mge_3 + \rho \boldsymbol{R}_B^W e_3 + \boldsymbol{R}_H^W \boldsymbol{F}_{ext} \tag{2.5}$$

$$\boldsymbol{I}_B\dot{\boldsymbol{\omega}}_B = -\boldsymbol{\omega}_B \times \boldsymbol{I}_B\boldsymbol{\omega}_B + \boldsymbol{\tau} + \boldsymbol{\tau}_{ext} \tag{2.6}$$

$$\dot{\boldsymbol{\Theta}}_W = \boldsymbol{T}(\boldsymbol{\Theta}_W)\boldsymbol{\omega}_B \tag{2.7}$$

where m is the mass of the quadrotor, $e_3 = [0\ \ 0\ \ 1]^T$ is the unitary vector along the $\boldsymbol{Z}$ axis, g is the gravity acceleration, $\ddot{\boldsymbol{p}}_W = [\ddot{x}\ \ddot{y}\ \ddot{z}]^T \in \mathbb{R}^3$ is the acceleration of quadrotor in $\mathcal{F}_W$, $\dot{\boldsymbol{\omega}}_B = [\dot{p}\ \ \dot{q}\ \ \dot{r}]^T \in \mathbb{R}^3$ is the angular acceleration of the quadrotor w.r.t. $\mathcal{F}_B$,

$$\boldsymbol{I}_B = \begin{bmatrix} I_{xx} & 0 & 0 \\ 0 & I_{yy} & 0 \\ 0 & 0 & I_{zz} \end{bmatrix} \tag{2.8}$$

is the diagonal inertia matrix of the quadrotor in body frame,

$$\boldsymbol{T}(\boldsymbol{\Theta}_W) = \begin{bmatrix} 1 & \sin(\phi)\tan(\theta) & \cos(\phi)\tan(\theta) \\ 0 & \cos(\phi) & -sin(\phi) \\ 0 & \sin(\phi)\sec(\theta) & \cos(\phi)\sec(\theta) \end{bmatrix} \in \mathbb{R}^{3\times 3} \tag{2.9}$$

is the standard transformation matrix from $\boldsymbol{\omega}_B$ to the Euler angle rates $\dot{\boldsymbol{\Theta}}_W \in \mathbb{R}^3$ and $\boldsymbol{F}_{ext} = [F_{ext_x}\ \ F_{ext_y}\ \ F_{ext_z}]^T \in \mathbb{R}^3$, $\boldsymbol{\tau}_{ext} = [\tau_{ext_x}\ \ \tau_{ext_y}\ \ \tau_{ext_z}]^T \in \mathbb{R}^3$ represent all additional forces in $\mathcal{F}_H$ and torques in $\mathcal{F}_B$ respectively acting on the quadrotor due to disturbances and external forces. Note that the gravity acceleration ge_3 does not need to be rotated from $\mathcal{F}_W$ to $\mathcal{F}_H$, since its only non-zero component is not affected by this rotation being $\vec{\boldsymbol{Z}}_H \parallel \vec{\boldsymbol{Z}}_W$. Note that it is also possible to express equation (2.5) in $\mathcal{F}_H$ as

$$m\ddot{\boldsymbol{p}}_H = -mge_3 + \rho \boldsymbol{R}_B^H e_3 + \boldsymbol{F}_{ext}. \tag{2.10}$$

While for control purpose the system defined by (2.5)-(2.7) in the $\mathcal{F}_W$ is used, in the estimator design it will be convenient to consider the system defined in $\mathcal{F}_H$ by equations (2.10), (2.6) and (2.7) because it reduces the number of system states required.

2.2.1 Model of the External Wrench

The external wrench $\boldsymbol{\Lambda}_{ext} = [\boldsymbol{F}_{ext}^T\ \ \boldsymbol{\tau}_{ext}^T]^T \in \mathbb{R}^6$ is defined as the stacked vector of the external forces in $\mathcal{F}_H$ and torques in $\mathcal{F}_B$ applied in the center of mass O_B. It represents the resultant of all forces and torques acting on the UAV which are not due to the nominal actuation or the nominal gravity force acting on the quadrotor.

Given this very generic definition, it is clear that $\boldsymbol{\Lambda}_{ext}$ may include a large variety of terms such as disturbances due to either external causes, as wind, or to mismatches between the nominal and real parameters of the model, for example a difference between the nominal and real mass of the UAV. The resultant of all these forces and torques is modeled as one disturbance wrench $\boldsymbol{\Lambda}_{dis}^B = [{\boldsymbol{F}_{dis}^B}^T\ \ \boldsymbol{\tau}_{dis}^T]^T \in \mathbb{R}^6$ expressed in $\mathcal{F}_B$ applied in the center of mass of the UAV. The disturbance force $\boldsymbol{F}_{dis}^B$ can also be expressed in the horizontal frame $\mathcal{F}_H$ by the use of an appropriate rotation matrix:

$$\boldsymbol{\Lambda}_{ext} = \boldsymbol{J}_H \boldsymbol{\Lambda}_{dis}^B = \begin{bmatrix} \boldsymbol{R}_B^H & \mathbf{0}_3 \\ \mathbf{0}_3 & \boldsymbol{I}_3 \end{bmatrix} \boldsymbol{\Lambda}_{dis}^B. \tag{2.11}$$

The objective therefore is to estimate this external wrench $\boldsymbol{\Lambda}_{ext}$ that is acting on the quadrotor dynamics.

2.3 External Wrench (Force/Torque) Observer

As mentioned earlier in Sec. 2.1.1, the external wrench ($\boldsymbol{\Lambda}_{ext}$) observer is a residual based estimator technique used earlier in manipulators for fault detection and isolation (Takakura *et al.*, 1989; De Luca and Mattone, 2003). Here, the FDI technique has been suitably modified for aerial vehicles to fit in the quadrotor model by considering all the $\boldsymbol{\Lambda}_{ext}$ which are not part of the quadrotor dynamics can be identified as part of the accumulated residual.

For its mathematical description, it is convenient to express the dynamical model of the quadrotor (2.10), (2.6) following the Lagrangian formulation (Khalil and Dombre, 2004):

$$\boldsymbol{M}\dot{\boldsymbol{\zeta}} + \boldsymbol{C}(\boldsymbol{\zeta})\boldsymbol{\zeta} + \boldsymbol{G} = \boldsymbol{\Lambda} + \boldsymbol{\Lambda}_{ext} \tag{2.12}$$

where

$$\boldsymbol{M} = \begin{bmatrix} m\boldsymbol{I}_3 & \mathbf{0}_3 \\ \mathbf{0}_3 & \boldsymbol{I}_B \end{bmatrix} \in \mathbb{R}^{6\times 6} \tag{2.13}$$

is the diagonal, positive definite inertial matrix. The matrix

$$\boldsymbol{C}(\boldsymbol{\zeta}) = \begin{bmatrix} \mathbf{0}_3 & \mathbf{0}_3 \\ \mathbf{0}_3 & \begin{bmatrix} 0 & I_{zz}r & -I_{yy}q \\ -I_{zz}r & 0 & I_{xx}p \\ I_{yy}q & -I_{xx}p & 0 \end{bmatrix} \end{bmatrix} \in \mathbb{R}^{6\times 6} \tag{2.14}$$

expresses the Coriolis and centrifugal terms, while $\boldsymbol{G}$ is the gravitational vector given by

$$\boldsymbol{G} = \begin{bmatrix} 0 & 0 & mg & 0 & 0 & 0 \end{bmatrix}^T \in \mathbb{R}^6 \tag{2.15}$$

and $\boldsymbol{\Lambda} = [(\rho \boldsymbol{R}_B^H e_3)^T \;\; \boldsymbol{\tau}^T]^T \in \mathbb{R}^6$ is the nominal wrench due to the control input which can be related to the rotational speed of the propellers Ω_1, Ω_2, Ω_3 and Ω_4 by (Bresciani,

2008):

$$\boldsymbol{\Lambda} = \begin{bmatrix} f_x \\ f_y \\ f_z \\ \tau_x \\ \tau_y \\ \tau_z \end{bmatrix} = \boldsymbol{J}_H \begin{bmatrix} 0 \\ 0 \\ b(\Omega_1^2+\Omega_2^2+\Omega_3^2+\Omega_4^2) \\ bl(\Omega_4^2-\Omega_2^2) \\ bl(\Omega_3^2-\Omega_1^2) \\ d(\Omega_2^2+\Omega_4^2-\Omega_1^2-\Omega_3^2) \end{bmatrix} \tag{2.16}$$

where b is the propeller lift coefficient, l is the length of the propeller arms from $\boldsymbol{O}_B$ and d is the propeller drag coefficient. From the definition of $\boldsymbol{J}_H$ in (2.11), it is also clearly visible in the wrench defined above in (2.16) that the translational dynamics are defined in terms of the horizontal frame $\mathcal{F}_H$ whereas the rotational dynamics are referred in terms of the body frame $\mathcal{F}_B$ of the quadrotor. Note that in the above equation (2.12), the terms related to the propeller dynamics are neglected because the static and viscous friction terms which are part of the standard Lagrangian formulation are usually small with respect to the other terms and are combined with the external disturbance.

The external wrench estimator is based on the idea of the generalized momenta $\boldsymbol{Q} = \boldsymbol{M}\boldsymbol{\zeta}$, for which it is possible to write the following first-order dynamic equation:

$$\dot{\boldsymbol{Q}} = \boldsymbol{\Lambda} + \boldsymbol{\Lambda}_{ext} + \boldsymbol{C}^T(\boldsymbol{\zeta})\boldsymbol{\zeta} - \boldsymbol{G} \tag{2.17}$$

which is obtained from (2.12). Let the residual vector $\boldsymbol{r} \in \mathbb{R}^6$ for the disturbance estimation of the quadrotor be defined as

$$\boldsymbol{r}(t) = \boldsymbol{K}_I\left(\boldsymbol{Q} - \int_0^t (\boldsymbol{\Lambda} + \boldsymbol{C}^T(\boldsymbol{\zeta})\boldsymbol{\zeta} - \boldsymbol{G} + \boldsymbol{r})dt\right) \tag{2.18}$$

where $\boldsymbol{K}_I \succ 0$ is a diagonal positive-definite gain matrix. For $\boldsymbol{r}(0) = 0$, the dynamic evolution of $\boldsymbol{r}$ satisfies,

$$\dot{\boldsymbol{r}} = \boldsymbol{K}_I(\boldsymbol{\Lambda}_{ext} - \boldsymbol{r}) \tag{2.19}$$

which is an exponentially stable linear system driven by the external disturbance wrench. For the implementation of (2.18) at every time instant only the measure of the current $\boldsymbol{\zeta}$ (i.e., velocity) and the knowledge of the commanded wrench $\boldsymbol{\Lambda}$ are required. Equation (2.19) shows that the dynamic evolution of '$\boldsymbol{r}$' has a stable first-order filter structure. Therefore the transfer function of each component of the residual vector takes the form

$$\frac{\boldsymbol{r}_I(s)}{\boldsymbol{\Lambda}_{ext,I}(s)} = \frac{\boldsymbol{K}_I}{s+\boldsymbol{K}_I}, \quad I = 1,\ldots,6 \tag{2.20}$$

which has a unitary gain. Therefore, for "sufficiently" large gains, the evolution of $\boldsymbol{r}_I(t)$

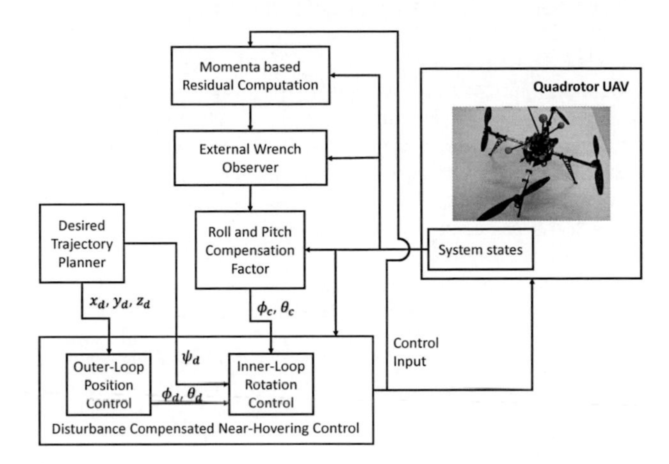

Figure 2.2: Block diagram of the system architecture showing the estimation and the control framework.

resembles $\boldsymbol{\Lambda}_{ext,I}(t)$ and the dynamic residual in (2.19) becomes

$$\boldsymbol{r} \simeq \boldsymbol{\Lambda}_{ext}. \tag{2.21}$$

Hence, the following is used as estimator of the external wrench:

$$\hat{\boldsymbol{\Lambda}}_{ext} = \boldsymbol{r}, \tag{2.22}$$

where in general the symbol $\hat{\circ}$ indicates the estimated value of a quantity $\circ$.

Remark 1. *The residual vector gives deeper knowledge about the disturbance components* $\boldsymbol{\Lambda}_{ext}$ *that are affecting the quadrotor dynamics. If a particular component of* $\boldsymbol{\Lambda}_{ext}$ *defined in (2.11) is zero then the scalar residual value corresponding to that component in (2.22) converges at zero. In presence of* $\boldsymbol{\Lambda}_{ext}$*, one or more residuals rise above the threshold corresponding to the external wrench.*

Remark 2. *The evolution of the system defined by equation (2.19) is dictated by the gain matrix* $\boldsymbol{K}_I$*. In particular, the larger the values of the gains, the faster and more accurately the residual in (2.18) will converge to the actual value. On the other hand, too large values of* $\boldsymbol{K}_I$ *will result in noisy (less precise) estimates. Hence, the gain matrix* $\boldsymbol{K}_I$ *must be tuned taking into account these aspects.*

2.4 Disturbance Compensator in Near-Hovering Control

The focus of this section is on improving a popular control law, proposed in (Michael *et al.*, 2010),(Lee *et al.*, 2013), for quadrotor UAVs to perform near-hovering flights and trajectory tracking, through the application of the external wrench estimator developed in Sec. 2.3. Therefore, the equations of the standard near-hovering controller are briefly recalled initially with the control law being defined in horizontal frame $\mathcal{F}_H$. Then, a feed-forward term to compensate the estimated wrench is derived, to show how the observed external wrench $\hat{\boldsymbol{\Lambda}}_{ext}$ can be integrated in the control scheme.

2.4.1 Standard near-hovering control

The near-hovering control law is based on the quadrotor dynamical equations which has a natural decoupling property, i.e., the attitude dynamics $\dot{\boldsymbol{\omega}}_B$ is independent from the translational dynamics $\ddot{\boldsymbol{p}}_W$, as seen from equations (2.5), (2.6), (2.7). The controller is designed such that it has an inner-outer loop structure. The position tracking slower outer loop is designed for the translational dynamics in (2.5) to drive $\boldsymbol{p}_W$ towards $\boldsymbol{p}_d$, while satisfying the thrust ρ and the attitude commands $\boldsymbol{\tau}$, whereas the attitude controller is designed as a faster inner loop for the rotational dynamics in (2.6) and (2.7) for the desired attitude (ϕ_d and θ_d) derived from the outer loop. Figure. 2.2 shows the control framework. Here $\boldsymbol{p}_d = [x_d \;\, y_d \;\, z_d]^T \in \mathbb{R}^3$ represents the desired position trajectory input given to the controller.

The goal of the controller therefore is to separately control the position $\boldsymbol{p}_W$ of the quadrotor to track a desired position $\boldsymbol{p}_d$ and the yaw angle ψ_d. Let the position error $\boldsymbol{p}_e$ be defined as

$$\boldsymbol{p}_e = \boldsymbol{p}_W - \boldsymbol{p}_d = \begin{bmatrix} e_x \\ e_y \\ e_z \end{bmatrix} = \begin{bmatrix} x - x_d \\ y - y_d \\ z - z_d \end{bmatrix}. \tag{2.23}$$

To design the position controller, expanding and rearranging the terms of the rotation matrix in (2.5) to get

$$m\ddot{z} = -mg + \rho\cos\phi\cos\theta + F_{ext_z} \tag{2.24}$$

for the Z-axis dynamics. Upon rearranging (2.24) for the value of commanded thrust ρ and excluding the external disturbance, it becomes

$$\rho = \frac{m}{\cos\phi\cos\theta}[g + \ddot{z}]. \tag{2.25}$$

Here F_{ext_z}, the external force in Z-direction, is neglected in the controller since it would be compensated as part of the compensation factor. Designing a PD controller for $\ddot{z}$

in (2.25), ρ becomes

$$\rho = \frac{m}{\cos\phi\cos\theta}[g + \ddot{z}_d + k_{d_z}(\dot{z}_d - \dot{z}) + k_{p_z}(z_d - z)] \tag{2.26}$$

where $k_{p_z} \in \mathbb{R}$ and $k_{d_z} \in \mathbb{R}$ are respectively the proportional and derivative gain of the PD controller to reach z_d. Note that (2.26) ensures stability as long as the quadrotor stays away from singularity conditions. Substituting (2.26) in the dynamic equations for $\ddot{x}$ and $\ddot{y}$ from the first two rows of (2.5), to have

$$m\begin{pmatrix} \ddot{x} \\ \ddot{y} \end{pmatrix} = \rho\,\Delta\begin{pmatrix} \sin\theta \\ \sin\phi \end{pmatrix} + \begin{pmatrix} F_{ext_x} \\ F_{ext_y} \end{pmatrix}. \tag{2.27}$$

When $\Delta = \left(\begin{smallmatrix} \cos\phi\cos\psi & \sin\psi \\ \cos\phi\sin\psi & -\cos\psi \end{smallmatrix}\right) \in \mathbb{R}^{2\times 2}$ is always analytically invertible, the error e_x and e_y are exponentially stable, if the attitude controller is able to roll and pitch with ϕ_d and θ_d given by

$$\begin{pmatrix} \sin\theta_d \\ \sin\phi_d \end{pmatrix} = \frac{m\Delta^{-1}}{\rho}\begin{pmatrix} \ddot{x}_d + k_{d_x}(\dot{x}_d - \dot{x}) + k_{p_x}(x_d - x) \\ \ddot{y}_d + k_{d_y}(\dot{y}_d - \dot{y}) + k_{p_y}(y_d - y) \end{pmatrix} \tag{2.28}$$

where k_{p_x}, k_{p_y} and k_{d_x}, k_{d_y} are the corresponding proportional and derivative gain of the PD controller to attain x_d and y_d respectively. Though equation (2.28) defines a nonlinear equation, in practice it is found to be working with a fast enough attitude control.

For the inner loop defining the attitude controller, clearly the ϕ_d and the θ_d are calculated at every time instant from the outer position controller loop as seen in (2.28) and the ψ_d is set as desired by the user. Differentiating (2.7), it is obtained

$$\ddot{\boldsymbol{\Theta}}_W = \boldsymbol{T}(\boldsymbol{\Theta}_W)\dot{\boldsymbol{\omega}}_B + \dot{\boldsymbol{T}}(\boldsymbol{\Theta}_W)\boldsymbol{\omega}_B. \tag{2.29}$$

Substituting $\dot{\boldsymbol{\omega}}_B$ from (2.6) in (2.29), it becomes

$$\ddot{\boldsymbol{\Theta}}_W = \boldsymbol{T}(\boldsymbol{\Theta}_W)\boldsymbol{I}_B^{-1}(-\boldsymbol{\omega}_B \times \boldsymbol{I}_B\boldsymbol{\omega}_B + \boldsymbol{\tau} + \boldsymbol{\tau}_{ext}) + \dot{\boldsymbol{T}}(\boldsymbol{\Theta}_W)\boldsymbol{\omega}_B. \tag{2.30}$$

Separating $\boldsymbol{\tau}$ from the above equation by rearranging (2.30), we get

$$\boldsymbol{\tau} = \boldsymbol{T}(\boldsymbol{\Theta}_W)^{-1}\boldsymbol{I}_B\left(\ddot{\boldsymbol{\Theta}}_W - \dot{\boldsymbol{T}}(\boldsymbol{\Theta}_W)\boldsymbol{\omega}_B\right) + \boldsymbol{\omega}_B \times \boldsymbol{I}_B\boldsymbol{\omega}_B - \boldsymbol{\tau}_{ext}. \tag{2.31}$$

Therefore the attitude regulation control for $\boldsymbol{\Theta}_d = \begin{bmatrix}\phi_d & \theta_d & \psi_d\end{bmatrix}^T \in \mathbb{R}^3$ can then be chosen as

$$\boldsymbol{\tau} = \boldsymbol{T}(\boldsymbol{\Theta}_W)^{-1}\boldsymbol{I}_B(k_{d_{\boldsymbol{\Theta}}}\dot{\boldsymbol{\Theta}} + k_{p_{\boldsymbol{\Theta}}}(\boldsymbol{\Theta}_d - \boldsymbol{\Theta})), \tag{2.32}$$

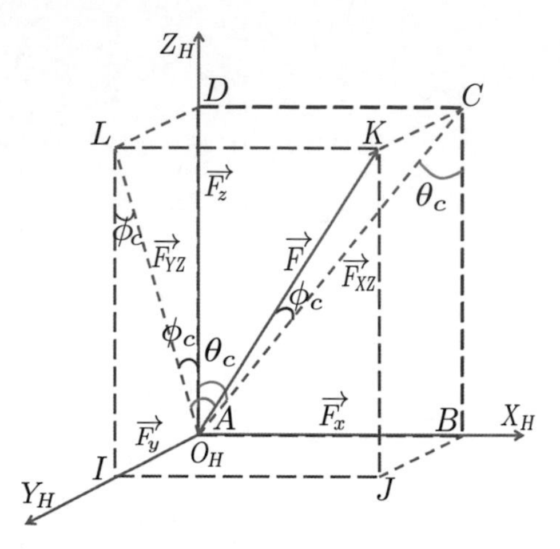

Figure 2.3: Graphical representation of the forces in the horizontal frame $\mathcal{F}_H$. The compensatory angles are roll ϕ_c (red) and pitch θ_c (green).

which will have a locally stable closed loop dynamics (Lee *et al.*, 2013) as,

$$\ddot{e}_{\boldsymbol{\Theta}}+\left[k_{d_{\boldsymbol{\Theta}}}+\boldsymbol{T}(\boldsymbol{\Theta}_W)\boldsymbol{I}_B^{-1}(-\boldsymbol{\omega}_B\times\boldsymbol{I}_B\boldsymbol{\omega}_B)-\dot{\boldsymbol{T}}(\boldsymbol{\Theta}_W)\boldsymbol{T}(\boldsymbol{\Theta}_W)^{-1}\right]\dot{e}_{\boldsymbol{\Theta}}+k_{p_{\boldsymbol{\Theta}}}e_{\boldsymbol{\Theta}}=\boldsymbol{T}(\boldsymbol{\Theta}_W)\boldsymbol{I}_B^{-1}\boldsymbol{\tau}_{ext}. \tag{2.33}$$

Here $e_{\boldsymbol{\Theta}}=\boldsymbol{\Theta}-\boldsymbol{\Theta}_d\in\mathbb{R}^3$ is the orientation error and $k_{p_{\boldsymbol{\Theta}}}\in\mathbb{R}^3$, $k_{d_{\boldsymbol{\Theta}}}\in\mathbb{R}^3$ are respectively the proportional and derivative gain to attain $\boldsymbol{\Theta}_d$.

For now it is evident that the controller summarized above is not the main concern or objective, but the ϕ_d and θ_d calculated at the end of the outer-loop (see Fig. 2.2) to be passed on to the fast inner-loop which results in the tracking of $\boldsymbol{p}_d$. How to possibly use the disturbance $\hat{\boldsymbol{\Lambda}}_{ext}$ estimated in Sec. 2.3 by utilizing disturbance compensation factor ϕ_c and θ_c that needs to be added to ϕ_d and θ_d respectively to follow any arbitrary trajectory when an external wrench $\boldsymbol{\Lambda}_{ext}$ is acting on the quadrotor is then the subject of discussion in Sec. 2.4.2.

2.4.2 Calculation of roll (ϕ_c) and pitch (θ_c) compensation

In a hovering condition, let $\boldsymbol{F}$ be the direction of the current 3D compensated force in the three positive axes $\{\vec{\boldsymbol{X}}_H,\vec{\boldsymbol{Y}}_H,\vec{\boldsymbol{Z}}_H\}$ of the horizontal frame $\mathcal{F}_H$. Therefore $\boldsymbol{F}$ is the compensation factor for external force disturbance $\boldsymbol{F}_{ext}=\left[F_{ext_x}\quad F_{ext_x}\quad F_{ext_x}\right]^T$ as shown in Fig. 2.3.

As seen in Fig. 2.3, $\boldsymbol{F}_{XZ}$ is the projection of $\boldsymbol{F}$ in the $\boldsymbol{X}_H\boldsymbol{Z}_H$ plane and $\boldsymbol{F}_{YZ}$ is the projection of $\boldsymbol{F}$ in the $\boldsymbol{Y}_H\boldsymbol{Z}_H$ plane. During hovering, the force $\{\boldsymbol{F}_x, \boldsymbol{F}_y, \boldsymbol{F}_z\}$ along the three principal axes $\{\vec{\boldsymbol{X}}_H, \vec{\boldsymbol{Y}}_H, \vec{\boldsymbol{Z}}_H\}$ respectively is given as

$$\begin{bmatrix} \boldsymbol{F}_x \\ \boldsymbol{F}_y \\ \boldsymbol{F}_z \end{bmatrix} = \begin{bmatrix} \boldsymbol{F}_{comp_x} \\ \boldsymbol{F}_{comp_y} \\ \boldsymbol{F}_{hover_z} + \boldsymbol{F}_{comp_z} \end{bmatrix}, \tag{2.34}$$

where $\boldsymbol{F}_{comp} = \begin{bmatrix} \boldsymbol{F}_{comp_x} & \boldsymbol{F}_{comp_y} & \boldsymbol{F}_{comp_z} \end{bmatrix}^T$ is the compensated for the external disturbance $\boldsymbol{\Lambda}_{ext}$ in $\mathcal{F}_H$ and $\boldsymbol{F}_{hover_z}$ is the force along $\boldsymbol{Z}_H$-axis for hovering. From the triangle ACD, it is clear that

$$\boldsymbol{F}_x = \boldsymbol{F}_{XZ} \sin\theta_c \tag{2.35}$$
$$\boldsymbol{F}_z = \boldsymbol{F}_{XZ} \cos\theta_c, \tag{2.36}$$

here θ_c is the pitch angle that is rotated around $\boldsymbol{Y}_H$ to create $\boldsymbol{F}_{comp_x}$. Similarly, from the triangle ADL, we get

$$\boldsymbol{F}_y = \boldsymbol{F}_{YZ} \sin\phi_c, \tag{2.37}$$

where ϕ_c is the roll angle that is rotated about $\boldsymbol{X}_H$ to create $\boldsymbol{F}_{comp_y}$. From the triangle AKC and the triangle AKL respectively, we have

$$\boldsymbol{F}_{XZ} = \boldsymbol{F} \cos\phi_c \tag{2.38}$$
$$\boldsymbol{F}_{YZ} = \boldsymbol{F} \cos\theta_c. \tag{2.39}$$

Now, substituting (2.38) in (2.36) and rearranging it is obtained

$$\boldsymbol{F} = \frac{\boldsymbol{F}_z}{\cos\phi_c \cos\theta_c}. \tag{2.40}$$

Substituting (2.38) in (2.35) and then using $\boldsymbol{F}$ from (2.40) to have

$$\boldsymbol{F}_x = \frac{\boldsymbol{F}_z \sin\theta_c}{\cos\theta_c}. \tag{2.41}$$

Similarly from (2.39) in (2.37) and then utilizing $\boldsymbol{F}$ from (2.40) to have

$$\boldsymbol{F}_y = \frac{\boldsymbol{F}_z \sin\phi_c}{\cos\phi_c}. \tag{2.42}$$

Therefore the roll and pitch compensation from (2.42) and (2.41) respectively

$$\phi_c = \arctan(\boldsymbol{F}_y / \boldsymbol{F}_z) \tag{2.43}$$

$$\theta_c = \arctan(\boldsymbol{F}_x/\boldsymbol{F}_z). \tag{2.44}$$

Note $\boldsymbol{F}_z = \boldsymbol{F}_{hover_z} + \boldsymbol{F}_{comp_z}$ as seen in (2.34). Here in the above equations (2.43) and (2.44), all the quantities are known as a result of the force/torque wrench observer at the end of Sec.2.3 and $\boldsymbol{F}_{hover_z}$ is the applied hovering thrust to balance the gravitational force along $\vec{Z}_H$.

2.5 Simulations and Analysis

We have performed physical simulations in order to validate: (*i*) the estimation of the external force and torque wrench acting on the quadrotor ($\hat{\boldsymbol{\Lambda}}_{ext}$) with a comparison with the known ground truth applied external wrench ($\boldsymbol{\Lambda}_{ext}$); (*ii*) the effect of the compensation factor from the $\hat{\boldsymbol{\Lambda}}_{ext}$.

The simulation environment and setup is based on Gazebo[2], a popular open source ROS-enabled simulator, which provides the dynamical simulation of the UAV and feedbacks the corresponding sensor readings (IMU, pose). The implementation of trajectory planner, observer, compensation factor, disturbance compensated near-hovering controller and generation of control inputs are executed in the Telekyb software (Grabe *et al.*, 2013) framework. Telekyb contains a collection of ROS nodes which provide hardware interfacing, estimation and control functionalities. Fig. 2.6 shows the block scheme of the simulation setup. The communication between the base station and Gazebo is achieved through ROS topics over IEEE 802.11 connection.

The physical simulation environment also faithfully produces noisy system state data for the position and orientation of the UAV. The state of the robot used for control and estimation purpose is estimated using an Extended Kalman filter which fuses the simulated noisy and biased IMU readings with the noisy measurements provided by a simulated pose sensor. In this context, we have considered two standard scenarios for UAV applications, namely, hovering and trajectory tracking, and two different external disturbance conditions: constant and variable wind.

2.5.1 Hovering with constant wind disturbance

In this simulation, initially the UAV is required to hover at a height of 1 m from the ground. During the hovering mode, a disturbance step is applied such that $\boldsymbol{F}_{ext} = \begin{bmatrix}1.38 & 1.38 & -0.42\end{bmatrix}^T N$ along $\{\vec{\boldsymbol{X}}_H, \vec{\boldsymbol{Y}}_H, \vec{\boldsymbol{Z}}_H\}$, the three principle axes of the horizontal frame, respectively. The external force/torque wrench observer proposed in Sec. 2.3 is used to estimate the applied disturbance. The estimated external forces $\hat{\boldsymbol{F}}_{ext}$ along the three axes shown in Figs. 2.4(a-c) validate the observer accuracy.

[2]http://gazebosim.org/

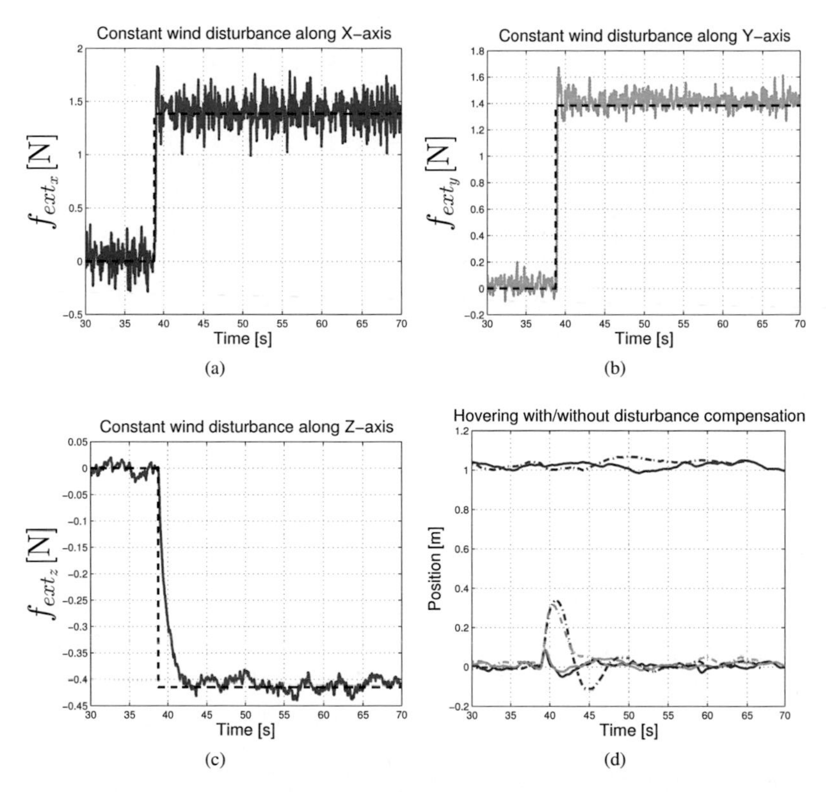

Figure 2.4: Results of the hovering simulation with constant wind. 2.4(a): Applied (black dashed line) and estimated (red solid line) disturbance along $\vec{X}_H$. 2.4(b): Applied (black dashed line) and estimated (green solid line) disturbance along $\vec{Y}_H$. 2.4(c): Applied (black dashed line) and estimated (blue solid line) disturbance along $\vec{Z}_H$. 2.4(d): Position $\boldsymbol{p}$ of the UAV: x (red), y (green) and z (blue) with disturbance-compensated near-hovering control (solid line) and standard near-hovering control (dashed line).

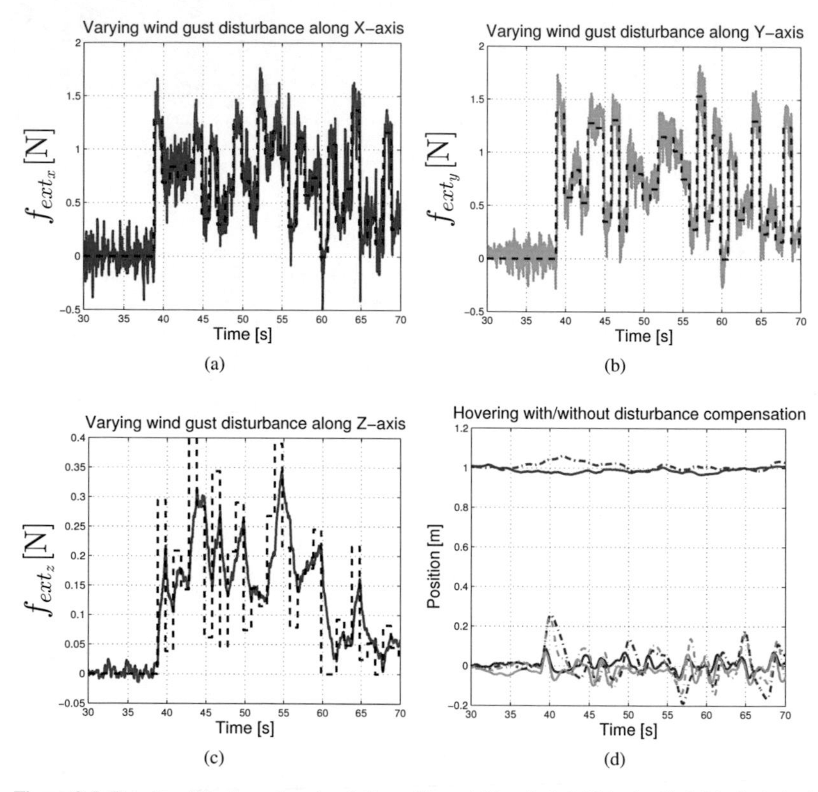

Figure 2.5: Results of the hovering simulation with variable wind. 2.5(a): Applied (black dashed line) and estimated (red solid line) disturbance along $\vec{X}_H$. 2.5(b): Applied (black dashed line) and estimated (green solid line) disturbance along $\vec{Y}_H$. 2.5(c): Applied (black dashed line) and estimated (blue solid line) disturbance along $\vec{Z}_H$. 2.5(d): Position $\boldsymbol{p}_W$ of the UAV: x (red), y (green) and z (blue) with disturbance-compensated near-hovering control (solid line) and standard near-hovering control (dashed line).

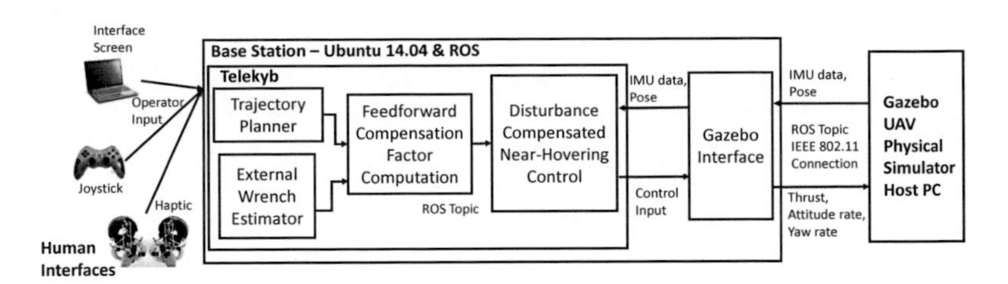

Figure 2.6: Block scheme of simulation setup

The next step is to assess the performance of the controller with the feedforward compensation factor. In Fig. 2.4(d) we show the results of a comparative analysis among the disturbance compensated near-hovering controller (red, green and blue solid lines respectively for the x, y and z coordinates of the UAV) and the standard near-hovering controller (red, green and blue dashed lines respectively for the x, y and z coordinates of the UAV). The settling time for the disturbance compensated controller is very short (less than $1s$) compared to the non-compensated near-hovering controller (more than $4s$). The compensation also drastically reduces the maximum position error from $\simeq 0.35m$ to $\simeq 0.05m$.

2.5.2 Hovering with varying wind disturbance

In the second hovering simulation, the quadrotor is set to hover at a height of 1 m from the ground as in Sec. 2.5.1 and then subjected to a variable wind along the three axes $\{\vec{X}_H, \vec{Y}_H, \vec{Z}_H\}$ with the strength of the disturbance varying between $0 \leq \boldsymbol{F}_{ext} \geq 1.5\ N$. The Figs. 2.5(a-c) shows the effective estimated $\hat{\boldsymbol{F}}_{ext}$ of the applied wind gust disturbance. In Fig. 2.5(d) the comparison of the not-compensated (dashed lines) and compensated (solid lines) near-hovering controllers show that the compensation factor effectively reduces the error in the hovering task also in case of variable wind gusts. As mentioned in Sec. 2.3, we obtain a faster and more accurate estimation convergence if a higher gain value $\boldsymbol{K}_I$ is chosen. However, this also results in noisy measurements which could result in unnecessary control action. Here in these simulations $\boldsymbol{K}_I = 10$ was fixed. This answers the reason for the slow estimate convergence as can be seen in Figs. 2.5(a-c).

2.5.3 Trajectory Tracking with constant wind disturbance

In this trajectory tracking task, the UAV is required initially to hover at a height of 1 m and then track a desired trajectory passing through the following points in space: $[0\ 0\ 1]^T$, $[2\ 0\ 2]^T$, $[2\ 2\ 1]^T$, $[0\ 2\ 2]^T$, $[0\ 0\ 1]^T$. At a certain time ($t = 38.5\ s$) during

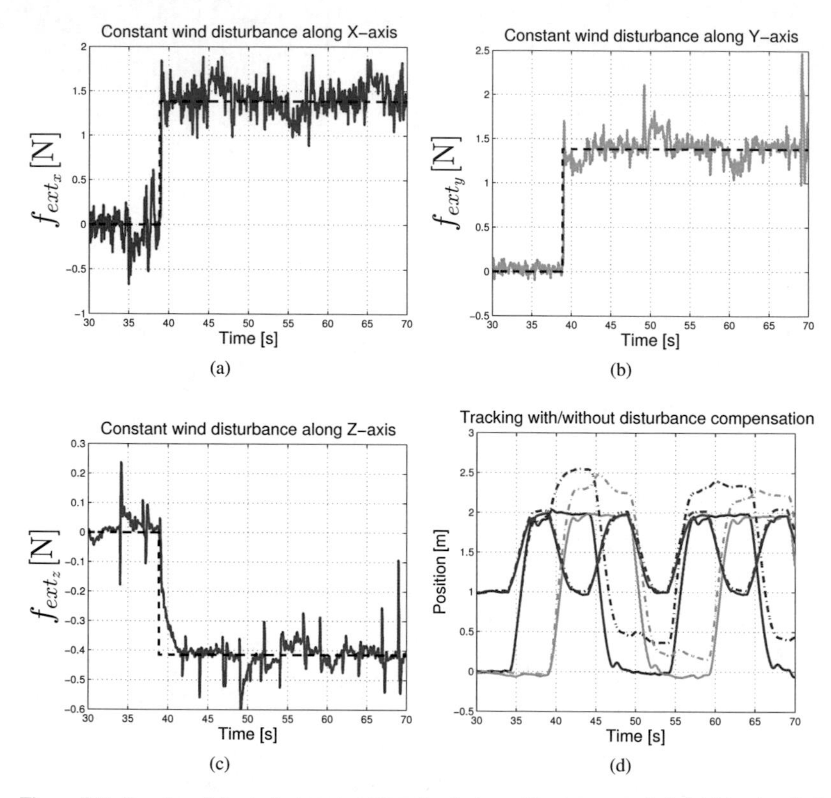

Figure 2.7: Results of the trajectory tracking simulation with constant wind. 2.7(a): Applied (black dashed line) and estimated (red solid line) disturbance along $\vec{\boldsymbol{X}}_H$. 2.7(b): Applied (black dashed line) and estimated (green solid line) disturbance along $\vec{\boldsymbol{Y}}_H$. 2.7(c): Applied (black dashed line) and estimated (blue solid line) disturbance along $\vec{\boldsymbol{Z}}_H$. 2.7(d): Position $\boldsymbol{p}_W$ of the UAV: x (red), y (green) and z (blue) with disturbance-compensated near-hovering control (solid line) and standard near-hovering control (dashed line).

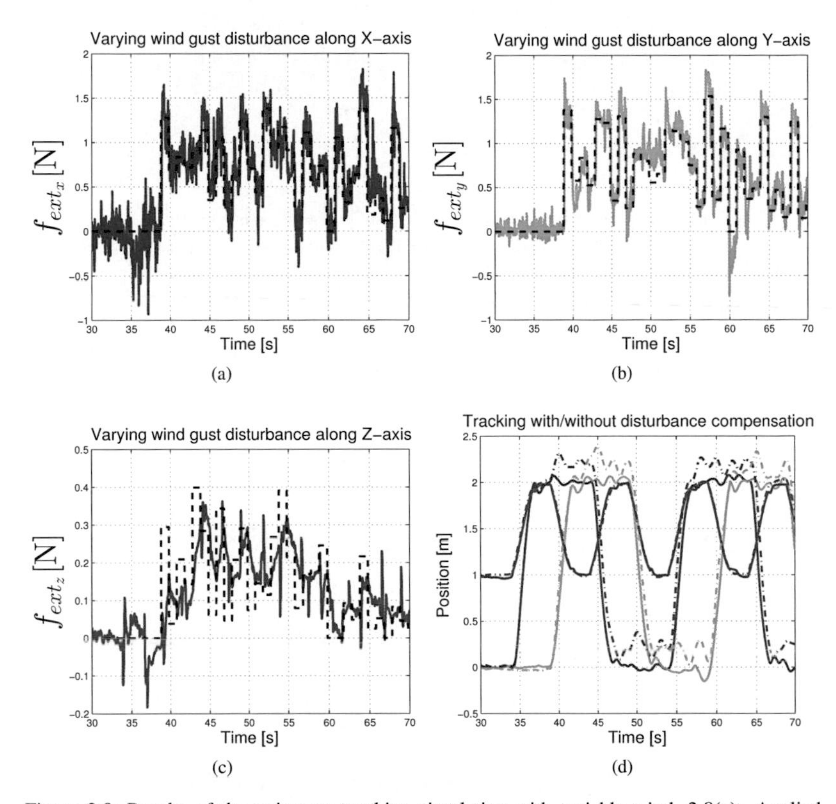

Figure 2.8: Results of the trajectory tracking simulation with variable wind. 2.8(a): Applied (black dashed line) and estimated (red solid line) disturbance along $\vec{X}_H$. 2.8(b): Applied (black dashed line) and estimated (green solid line) disturbance along $\vec{Y}_H$. 2.8(c): Applied (black dashed line) and estimated (blue solid line) disturbance along $\vec{Z}_H$. 2.8(d): Position $\boldsymbol{p}_W$ of the UAV: x (red), y (green) and z (blue) with disturbance-compensated near-hovering control (solid line) and standard near-hovering control (dashed line).

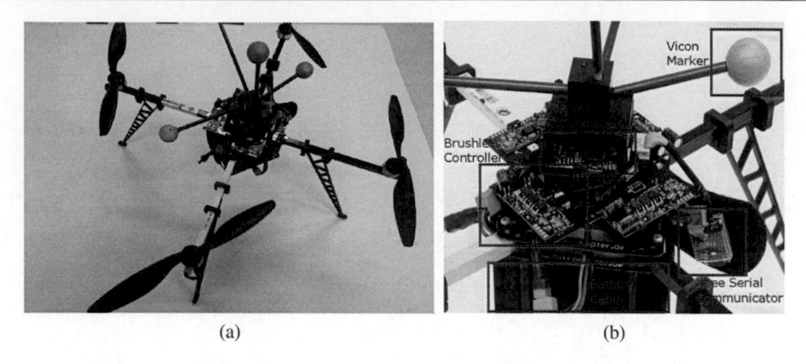

(a) (b)

Figure 2.9: Quadrotor experimental setup. 2.9(a): MK-Quadro from mikrocopter. 2.9(b): Hardware components in MK-Quadro consisting of flight controller, brushless controller, Xbee serial communicator, battery cabin and vicon markers.

the trajectory tracking, the UAV is subjected to a constant wind along the three principle axes of the horizontal frame $\{\vec{\boldsymbol{X}}_H, \vec{\boldsymbol{Y}}_H, \vec{\boldsymbol{Z}}_H\}$ respectively with disturbance of $\boldsymbol{F}_{ext} = \begin{bmatrix} 1.38 & 1.38 & -0.42 \end{bmatrix}^T N$. Figs. 2.7(a-c) show the $\hat{\boldsymbol{F}}_{ext}$ estimated during the trajectory tracking. Fig. 2.7(d) shows that without the disturbance-compensation factor, the standard near-hovering controller (dashed lines) produces an error $> 0.5\ m$ whereas the disturbance-compensated controller produces an error $< 0.05\ m$, proving the robustness of the disturbance estimation and compensation in trajectory tracking applications.

2.5.4 Trajectory Tracking with varying wind disturbance

A similar task of UAV hovering at 1 m height and tracking the way-points as the one in Sec. 2.5.3 is performed but with the application of variable wind. Figs. 2.8(a-c) show the applied and estimated disturbances. It is observed that $\hat{\boldsymbol{F}}_{ext_x}$ and $\hat{\boldsymbol{F}}_{ext_y}$ converge faster to the applied external wrench whereas $\hat{\boldsymbol{F}}_{ext_z}$ is slower. This is again attributed to the effect of the gain parameter $\boldsymbol{K}_I$ in the observer design. Fig. 2.8(d) shows that without the disturbance compensation factor (ϕ_c, θ_c in (2.43) and (2.44) respectively), the standard near-hovering controller produces a tracking error $> 0.4m$ whereas the disturbance-compensated controller produces a tracking error $< 0.05m$. This further confirms the robustness of the disturbance estimator and compensation in trajectory tracking applications.

Parameter	Description	Value	Unit
m	mass of the UAV	1	Kg
g	gravity acceleration	9.81	m/s^2
I_{xx}	inertia along X-axis	0.011549	$Kg.m^2$
I_{yy}	inertia along Y-axis	0.011368	$Kg.m^2$
I_{zz}	inertia along Z-axis	0.019444	$Kg.m^2$
b	lift coefficient	$1.6073 * 10^{-5}$	N/Ω^2
d	drag coefficient	$2.7988 * 10^{-7}$	Nm/Ω^2
$\boldsymbol{K}_I$	observer gain	10	-

Table 2.1: Experimental parameters

2.6 Experimental Validation

Experimental Setup

In order to validate the force/torque residual based wrench estimator for a real-time UAV application, experiments were conducted in a laboratory set-up. The UAV platform employed in this experiment is a *MK-Quadro* quadrotor from MikroKopter[3] as seen in Fig. 2.9(a). The standard UAV system parameters and the experimental parameters used are given in Table. 2.1. The MK-Quadro consists of 4 propeller arms, each one equipped with a motor controller, a brushless motor and a 10 inch propeller. The MK-Quadro has an onboard 8-bit microcontroller, which is used to perform the low-level control, transferring the control commands from the high-level controller to the motor controller. The microcontroller board includes an inertial measurement unit (IMU) composed of two 3-axis analog sensors: an accelerometer with measurement range of ± 2 g and a gyroscope with measurement range ± 300 deg/s, both read with a 10-bit analog to digital converter. The board communicates with the brushless motor controllers through a standard I^{2}C bus.

The standard MikroKopter firmware has been replaced with our own software that allows us to control the robot through a serial XBee channel operating at 115 200 Bd baud rate. The command sent to the microcontroller at $\sim$120 Hz, consist of the setpoints for the brushless motors, which are computed on an offboard desktop PC. The platform is powered by a 2600 mAh LiPo battery that provides approximately 10 min of flight.

The main control and estimation algorithms are performed in the base station, which is a ROS enabled Ubuntu 14.04 PC, with the Telekyb software framework (Grabe *et al.*, 2013). The state of the UAV in the $10m \times 10m$ flying arena is provided at 120 Hz by a motion capture system (VICON), which is also used to collect ground truth data. Fig. 2.9(b) shows the different hardware parts of the experimental setup.

All the experiments were conducted with the quadrotor in *hovering* mode and sub-

[3] http://www.mikrokopter.de/

jected to external forces and torques. In order to subject the UAV to known external forces and torques on the three axes, known suspended weights which produces equivalent force along $\boldsymbol{Z}$-axis were used. This generated force along $\boldsymbol{Z}$-axis were then appropriately transferred to the other axes as required by means of a *rope-pulley setup* as shown in Fig. 2.10. Frictionless pulleys were used so that losses due to friction are negligible, and a very light-weight rope was used to connect the weights to the quadrotor. With this setup, we have performed two separate experiments to test separately the force and torque estimations.

2.6.1 Force Estimation Experiment

The aim of this experiment is to estimate the external forces applied on the CoM of the quadrotor while it is in hovering. The forces were applied in steps, to test different forces acting on the quadrotor. As shown in Fig. 2.11, at the time instants $35s$, $46s$ and $57s$ respectively forces of $0.85N$, $2.17N$ and $4.35N$ were applied along $\vec{\boldsymbol{Z}}_H$ by adding weights of $87g$, $221g$ and $443g$. The estimated external force along the $\vec{\boldsymbol{Z}}_H$ axis, shown in blue, is correctly estimated by the observer.

At time $69s$ a weight of $222g$ acting along the $\vec{\boldsymbol{Z}}_H$ axis was removed and then at times $97s$, $107s$ and $119s$ respectively, forces of $0.22N$, $1.19N$ and $2.53N$ were added along a vector laying on the $\boldsymbol{X}_H\boldsymbol{Y}_H$ plane and having an azimuth of -60°. These forces were successively removed in the same order in which they were added. As shown in Fig. 2.11 the external forces $\boldsymbol{F}_{ext}$ were correctly estimated for F_{ext_x} (red), F_{ext_y} (green) and F_{ext_z} (blue). The weights added in the rope-pulley setup for the force experiment can be seen in Fig. 2.10(a-d, g, h).

2.6.2 Torque Estimation Experiment

Like the force estimation mentioned in Sec. 2.6.1, a similar experiment was conducted for the torque estimation with the same setup. In the first torque experiment the weights were suspended at the end of the arm of the quadrotor along the $-\vec{\boldsymbol{Y}}_H$. This creates a torque w.r.t. to $\vec{\boldsymbol{X}}_H$ namely τ_{ext_x}. During the experiment, a weight of $27g$ was suspended at a distance of $37.5cm$ from $\boldsymbol{O}_H$ in the $-\vec{\boldsymbol{Y}}_H$ axis as can be seen in Fig. 2.10(f, i). As shown in Fig. 2.12, a torque of $0.099Nm$ was applied. The Figure also shows the estimated torque τ_{ext_x} (red). A small constant τ_{ext_y} (green) is also estimated during the whole experiment, and it is due to a non perfect balance of the weights on the quadrotor.

In the second torque experiment, a force was applied through our rope-pulleys setup in order to create a torque τ_{ext_z} around the $\vec{\boldsymbol{Z}}_H$ axis. Similarly to the first experiment, a weight of $27g$ was suspended at a distance of $37.5cm$ from $\boldsymbol{O}_H$ as can be seen in Fig. 2.10(e, j). As we show in Fig. 2.13, τ_{ext_z} of $0.099Nm$ was correctly estimated (blue in Fig. 2.13). Overall, these experiments prove the effectiveness of the external torque estimator.

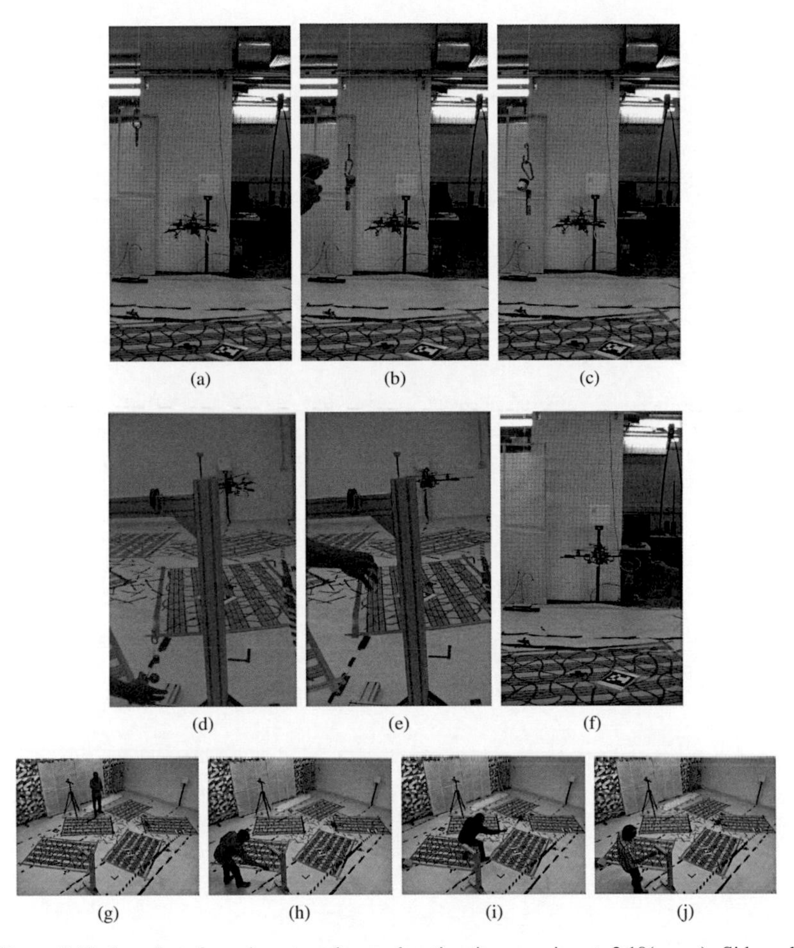

Figure 2.10: Snapshots from the external wrench estimation experiment. 2.10(a-c, g): Side and top view of force experiment along Z-axis. Known mass added at one end of the rope which is converted as force on the other end attached to the UAV through a rope-pulley setup. 2.10(d, h): Side and top view of force experiment along X and Y-axis. 2.10(f, i): Side and top view of torque experiment w.r.t. Y-axis. Known mass is suspended at the end of the arm to create a roll torque. 2.10(e, j): Side and top view of torque experiment w.r.t. Z-axis. Known mass is suspended at the one end of the rope whose other end is attached at the end of arm through a pulley to create a yaw torque.

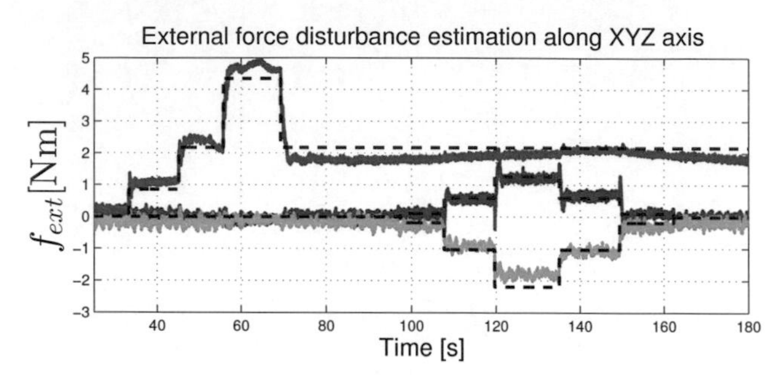

Figure 2.11: Results of the experiment with the external force disturbance along $\vec{X}_H$, $\vec{Y}_H$ and $\vec{Z}_H$. Applied disturbance (black dashed lines), estimated disturbance $\hat{F}_{ext_x}$ along $\vec{X}_H$ (red solid line), estimated disturbance $\hat{F}_{ext_y}$ along $\vec{Y}_H$ (green solid line) and estimated disturbance $\hat{F}_{ext_z}$ along $\vec{Z}_H$ (blue solid line).

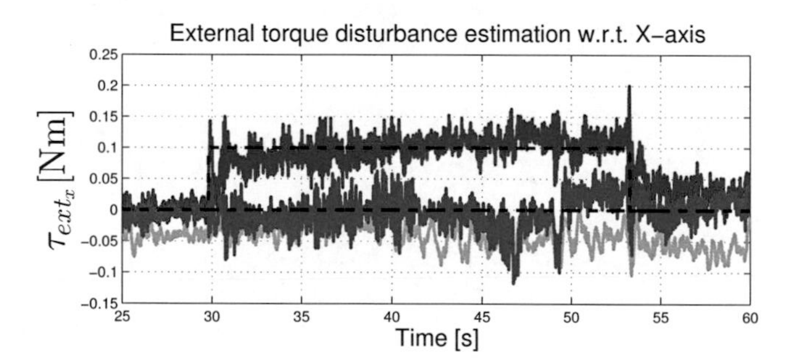

Figure 2.12: Results of the experiment with the external torque disturbance around $\vec{X}_H$. Applied disturbance (black dashed lines), estimated disturbance $\hat{\tau}_{ext_x}$ (red solid line), estimated disturbance $\hat{\tau}_{ext_y}$ (green solid line) and estimated disturbance $\hat{\tau}_{ext_z}$ (blue solid line).

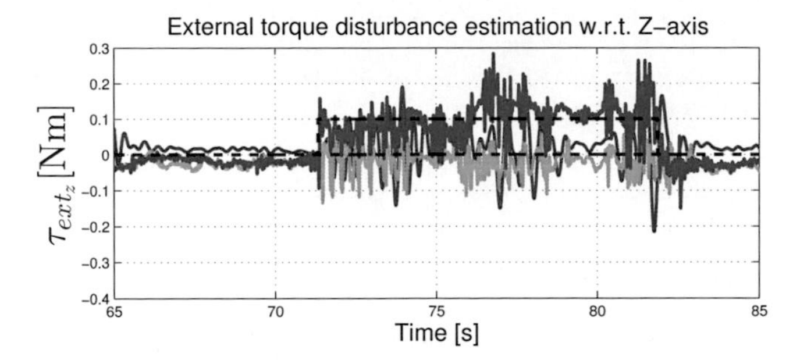

Figure 2.13: Results of the experiment with the external torque disturbance around $\vec{Z}_H$. Applied disturbance (black dashed lines), estimated disturbance $\hat{\tau}_{ext_x}$ (red solid line), estimated disturbance $\hat{\tau}_{ext_y}$ (green solid line) and estimated disturbance $\hat{\tau}_{ext_z}$ (blue solid line).

2.7 Discussions and Possible Extensions

Summarizing, the following results have been presented:

1. It was introduced the system dynamics of the standard quadrotor UAV (Sec. 2.2).

2. It was proposed an observer for external force/torque wrenches that act as disturbances on the quadrotor (Sec. 2.3). The estimator is based on the residual momenta based fault detection and identification technique. The dynamic evolution of the residual has a stable first-order filter structure, which is an exponentially stable linear system driven by the external wrenches.

3. It was computed a feedforward disturbance compensation factor utilizing the estimated external wrench. This factor which comprises of the roll compensator (ϕ_c) and pitch compensator (θ_c) can be applied to any standard controller as a feedforward term (see Sec. 2.4.2) to compensate for any deviation in trajectory tracking due to the external wrench. We have showed how this could be utilized in the standard near-hovering controller which in mostly commonly used with quadotor UAVs.

4. The external wrench observer is validated extensively both through physical simulations and experiments. The results confirms (i) the convergence of the observer with the actual external wrench and (ii) the wrench estimation approach is suitable for real-time application on aerial vehicles since there is no significant delay.

The observed force and torques are dependent on gain value $\boldsymbol{K}_I$ for the convergence speed as mentioned in Remark. 2. Therefore $\boldsymbol{K}_I$ should be carefully calibrated depending on the application. The main advantage of the estimator is that, this approach doesn't introduce additional mass in to the UAV mechanical setup. Therefore there is no compromise in the already existing flight time.

Since this estimation approach is proved applicable to an aerial vehicle, there are wide range of application that becomes possible

1. the very first one would be to improve the performance capability of the already existing controllers for aggressive maneuvers by introducing the observer in the control design;

2. this approach estimates all the external wrench in the center of mass. This could be very well extended to know the exact location from where the external wrench was detected by adding additional sensors. This would be very much useful when this observer is employed for an interaction application. This application scenario is investigated as part of this thesis in Chapter. 3

Chapter 3

Novel Architecture for Human-UAV Physical Interaction

Applications with UAVs are increasing day by day from simple surveillance towards UAV interaction with the environment. Manipulators are being equipped on aerial vehicles and successfully used in interaction tasks in space not reachable and/or safe for humans. Technology is moving in the direction of humans and UAVs physically interacting to accomplish tasks. But, *Is it possible and safe to have Human-UAV physical interaction and co-existence? How should be the hardware setup for such interaction? How is it possible to exchange forces and torques with UAVs? How will the UAV understand and separate the interaction wrench from disturbances? How should the control framework be organized in such a scenario?*

In order to answer to these questions, in this chapter we propose novel UAV hardware designs for human-UAV physical interaction. We also propose methodologies for interaction wrenches estimation and control. With this futuristic vision of humans and UAVs sharing the same workspace, we discuss the current technological limitations with respect to current UAV platforms, hardware architectures and software frameworks.

The discussion presented in this chapter is based upon the work that I have done under the supervision of Dr. Paolo Stegagno and is to appear in Rajappa *et al.* (2017a).

3.1 Introduction

Quadrotor UAVs are very popular for research purposes due to their ability of vertical takeoff and landing, unsophisticated mechanical design and relatively simple system dynamics. With the advancements in computer vision and control techniques, quadrotors are now able to achieve versatile tasks, like autonomous navigation and mapping (Heng *et al.*, 2011; Fraundorfer *et al.*, 2012), search and rescue (Mueggler *et al.*, 2014), goods transportation (Palunko *et al.*, 2012), construction (Augugliaro *et al.*, 2014), aerial acrobatics (Lupashin *et al.*, 2010; Ritz *et al.*, 2012; Mellinger *et al.*, 2012), grasping (Pounds *et al.*, 2011; Lindsey *et al.*, 2012) and aerial manipulation (Orsag *et al.*, 2013; Lippiello and Ruggiero, 2012; Gioioso *et al.*, 2014b).

Although most of current applications involves flying in areas that are hardly accessible for humans, the increasing number of commercial platforms that have hit the market and are available to companies and the great public will eventually call for always growing integration of quadrotors inside human-populated areas and facilities. Some form of interaction between the UAVs and the humans will become necessary in order to exchange information and many foreseeable applications will also involve some form of physical contact and force exchange (i.e.: physical interaction). For example, deploying tools to a worker exchanging forces at pick up or kinesthetic trajectory teaching to allow untrained users setting up UAV systems. Moreover, detecting mutually applied forces and implementing behaviors that increase the safety of nearby humans is necessary to let humans and UAVs share the same space.

However, most works in Human Robot Interaction (HRI) involving UAVs are limited to recognizing voice commands, hand gestures, body and face poses, since aerial vehicles are still classified as dangerous and not safe for human-robot physical interaction (HRPI). Among many reasons, the most important are: (i) the lack of an established hardware and software framework for UAV-HRPI; (ii) the lack of a proper interacting surface. The goal therefore is to fill this gap by developing the technology to perform UAV-HRPI, implementing a setup that allows exchanges of forces between humans and UAVs and the methodologies to detect, interpret and react to such forces. Albeit safety is not explicitly considered in this work, both the presence of an interaction surface interposed between the user and the propellers, as well as the behavior chosen for the UAV in case of a physical interaction (the motion in the opposite direction with respect to the interacting force) provide some degree of safety for the user.

3.1.1 Related works

In the context of aerial robotics, HRI has been explored mostly by considering either intermediary physical interfaces as monitors, joysticks and haptic devices, or visual and auditory sensory channels. Quigley *et al.* (2004) discussed different paradigms for human-UAV interfacing, with a detailed qualitative as well as quantitative performance analysis. The use of haptic interfaces for HRI with UAVs have been explored by Lee *et al.* (2013).

Concerning direct interaction with the UAV, Ng and Sharlin (2011) studied a gesture-based interactive scheme to communicate with the UAV based on a multimodal falconry metaphor. Lichtenstern *et al.* (2012) developed a command set for multi-robot systems using hand gestures recognized through an RGB-D sensor mounted on one of the UAVs. Pfeil *et al.* (2013) explored the possibility to use the upper body to communicate with a UAV. Naseer *et al.* (2013) used an active RGB-D sensor with vision based ego-motion cancellation to recognize and respond to hand gestures for high level tasks such as filming, landing, etc. Another approach by Monajjemi *et al.* (2013) uses a front facing camera for face tracking and gesture recognition to command robot teams. Sanna *et al.* (2013) implemented a visual odometry algorithm based on Kinect sensor to allow the platform to navigate as well as recognize gestures and body postures. In Nagi *et al.*

Figure 3.1: Our quadrotor setup for Human-UAV physical interaction.

(2014), a machine vision technique is used to control UAVs using face poses and hand gestures. Szafir *et al.* (2015) explored the visual communication of the directionality and intend of the aerial vehicle to the human user by means of coordinated lighting of set of Light Emitting Diodes (LEDs). Cauchard *et al.* (2015) conducted an evaluation study through Wizard-of-Oz elicitation about how users would naturally interact with drones, suggesting that it could be with voices, gestures or both.

Although rich, the above literature did not address UAV-HRPI. However, in recent years many works developed enabling technologies that can be employed for UAV-HRPI. Some works focused on new mechanical designs. For example, Briod *et al.* (2014) developed the gimball, which allows stable UAV flight even in the event of collisions thanks to the outer protective frame. With respect to this work, we also try to address the UAV-HRPI by means of designing a suitable and versatile software architecture which can be particularized for different tasks. Other works focused on the problem of making and keeping contact with some object in the environment. Fumagalli *et al.* (2012), designed an attitude controller along with a passivity-based controller for contact inspection using aerial manipulation. This methodology has been further expanded by using an impedance-force control hybrid architecture by Scholten *et al.* (2013), and modified impedance control by Fumagalli and Carloni (2013). A force control approach with an external feedforward signal has been used by Albers *et al.* (2010).

When dealing with UAV-HRPI, it is important to estimate and characterize the external wrench (force and torque) acting on the UAV. Augugliaro and D'Andrea (2013), proposed an unscented Kalman Filter in order to estimate the external wrench. To the best of our knowledge, this is also the only work which explicitly considers human-UAV physical interaction by proposing the use of an admittance controller. Bellens *et al.* (2012), investigated the problem of estimating the external wrench in the context of a hybrid pose/wrench control for a contact maintenance task. A force sensor is used as an estimator by Nguyen and Lee (2013). An alternative Lyapunov-based nonlinear observer for estimating the external wrench has been proposed by Yüksel *et al.* (2014) and numerically validated. An external wrench estimation method based on the generalized momenta developed by Magrini *et al.* (2014) for an arm manipulator, has been employed in the estimation of the external wrench acting on an aerial vehicle (Tomic and Haddadin, 2015; Ruggiero *et al.*, 2014). In particular, Tomic and Haddadin (2015), used this approach for collision detection, where they further separate the different contact forces from the aerodynamic disturbances based on the natural contact frequency characteristics. To the best of our knowledge, this work is the only one which tries to give a more rich characterization of the external wrench by considering simultaneously disturbances and contact forces.

From the control perspective, in UAV-HRPI it is important to ensure stable flight and disturbance rejection through robust control techniques. Controllers in literature include adaptive control methods (Roberts and Tayebi, 2011; Palunko *et al.*, 2012; Antonelli *et al.*, 2013), model predictive control approaches (Alexis *et al.*, 2011; Raffo *et al.*, 2010), backstepping, sliding mode (Bouabdallah and Siegwart, 2005, 2007), and super twisting controllers (Derafa *et al.*, 2012; Rajappa *et al.*, 2016).

3.1.2 Methodologies

The main objective of this work is to develop a framework and platform which allows UAV-HRPI. In order to achieve this objective,

1. it is employed the momenta-based external wrench observer developed and validated in Chapter. 2;
2. using a custom-designed sensor ring, it is proceeded to characterize the external wrench by separating human interaction forces from external disturbances through the formulation of a quadratic optimization problem;
3. it is then implemented an admittance control framework where it is changed the desired trajectory based on the interaction wrenches by manipulating the physical properties of the aerial vehicle considering them as a mass-spring-damper system. This control framework also includes a disturbance compensated geometric controller (Lee *et al.*, 2010) for tracking the resulting trajectory;
4. in addition, it is provided an extensive experimentation in which it is showed how varying only one parameter of the admittance control is enough to provide a wide

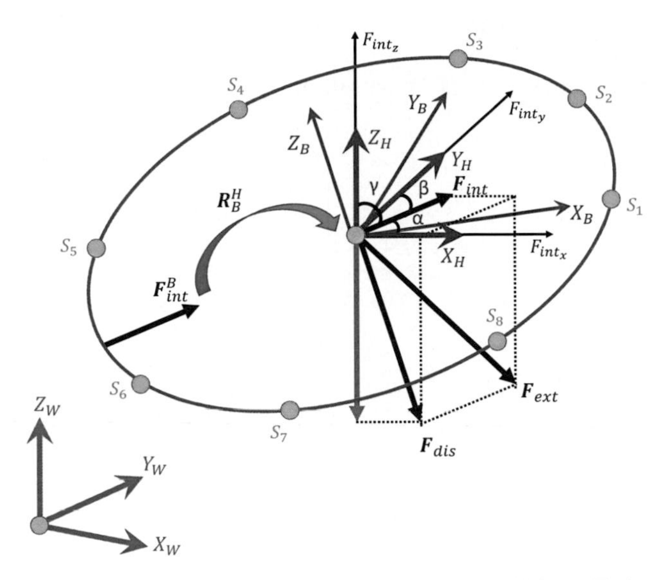

Figure 3.2: Visualization of the main frames and quantities: the world frame $\mathcal{F}_W$ in green; the body frame $\mathcal{F}_B$ in blue; the horizontal frame $\mathcal{F}_H$ in red; external $\boldsymbol{F}_{ext}$, disturbance $\boldsymbol{F}_{dis}$ and interaction $\boldsymbol{F}_{int}$ forces in black; green circles represents the sensors.

range of behaviors suitable for several different applications. To the best of knowledge, this is the first work to show this concept applied to UAVs.

In the context of UAV-HRPI, the contribution of this work are,

1. a novel hardware design for UAV-HRPI, including a sensor ring which provides an interaction surface and useful data for the characterization of the external wrench;

2. an easy to implement and novel methodology for the separation of the forces/-torques applied by an interacting human from generic disturbances;

3. the general architecture of the estimation and control framework which allows a human to provide intuitive force command to the UAV while the disturbance are rejected by the controller.

3.2 Problem Setting

3.2.1 Preliminary System Descriptions

The quadrotor system dynamics follow the notations and equations as introduced earlier in Sec. 2.2 of Chapter. 2.

The different frame definition: the world frame $\mathcal{F}_W$, the body frame $\mathcal{F}_B$ and the horizontal frame $\mathcal{F}_H$ follows the same philosophy as seen in Fig. 2.1. The basic quadrotor states $\boldsymbol{\xi}_W$ defined in $\mathcal{F}_W$ is as in (2.1) and the state $\boldsymbol{\xi}_H$ of the UAV in $\mathcal{F}_H$ which is utilized in the wrench estimation is expressed in (2.3). The generalized velocity vector states $\boldsymbol{\zeta}$ are expressed as defined in (2.4). The dynamical model of the quadrotor expressed using the Newton-Euler formulation in $\mathcal{F}_W$ is given by (2.5)-(2.7). The translational dynamics in the horizontal frame $\mathcal{F}_H$ is given by (2.10).

As mentioned earlier for control purpose we will use the system defined by (2.5)-(2.7), whereas in the estimator design it will be convenient to consider the system defined by equations (2.10), (2.6) and (2.7). With the system description defined similarly as in Chapter. 2, how to suitably define an external wrench model for $\boldsymbol{\Lambda}_{ext}$ which includes the human physical interaction wrenches and how to efficiently control them is the subject of discussion in the following sections.

3.2.2 Extended Model of the External Wrench

Let the extended external wrench $\boldsymbol{\Lambda}_{ext_{ex}} = [\boldsymbol{F}_{ext}^T \ \ \boldsymbol{\tau}_{ext}^T]^T \in \mathbb{R}^6$ be defined as the stacked vector of the external forces in $\mathcal{F}_H$ and torques in $\mathcal{F}_B$ applied in the center of mass O_B. $\boldsymbol{\Lambda}_{ext_{ex}}$ represents the resultant of all forces and torques acting on the UAV which includes not only the external disturbances, model mismatches (as defined in Sec. 2.2.1) but also the human physical interaction wrenches.

The components of $\boldsymbol{\Lambda}_{ext_{ex}}$ can be separated into two main categories. In the first category, we consider all disturbances due to either external causes, as wind, or to mismatches between the nominal and real parameters of the model. The resultant of all these forces and torques is modeled as one disturbance wrench $\boldsymbol{\Lambda}_{dis}^B = [\boldsymbol{F}_{dis}^{B\ T} \ \ \boldsymbol{\tau}_{dis}^T]^T \in \mathbb{R}^6$ expressed in $\mathcal{F}_B$ applied in the center of mass of the UAV. The disturbance force $\boldsymbol{F}_{dis}^B$ can also be expressed in the horizontal frame $\mathcal{F}_H$ by the use of an appropriate rotation matrix:

$$\boldsymbol{\Lambda}_{dis} = \boldsymbol{J}_{H_{dis}} \boldsymbol{\Lambda}_{dis}^B = \begin{bmatrix} \boldsymbol{R}_B^H & \boldsymbol{0}_3 \\ \boldsymbol{0}_3 & \boldsymbol{I}_3 \end{bmatrix} \boldsymbol{\Lambda}_{dis}^B . \tag{3.1}$$

The forces and torques resulting from physical interaction with humans lie in the second category. In general, one or more interacting humans apply q independent wrenches $\boldsymbol{\Lambda}_{int_i}^B = [\boldsymbol{F}_{int_i}^{B\ T} \ \ \boldsymbol{\tau}_{int_i}^T]^T \in \mathbb{R}^6, i = 1, \ldots, q$, where $\boldsymbol{F}_{int_i}^B$ and $\boldsymbol{\tau}_{int_i}$ are both expressed in $\mathcal{F}_B$. The application points $\boldsymbol{p}_i, i = 1, \ldots, q$ expressed in $\mathcal{F}_B$ are in general all different from each other. In order to introduce those wrenches in equations (2.5) and (2.6), we must first express the forces in $\mathcal{F}_H$ as $\boldsymbol{F}^H_{int_i} = R^H_B \boldsymbol{F}^B_{int_i}$. Then we need to compute their effect on the center of mass, since they cause additional torques $-[\boldsymbol{R}^H_B \boldsymbol{p}_i]_\wedge$ where $[\cdot]_\wedge$ is the map from $\mathbb{R}^3$ to the skew-symmetric matrix

$$\left[\begin{bmatrix} x \\ y \\ z \end{bmatrix}\right]_\wedge = \begin{bmatrix} 0 & -z & y \\ z & 0 & -x \\ -y & x & 0 \end{bmatrix}. \tag{3.2}$$

Hence, the generic wrench $\boldsymbol{\Lambda}^B_{int_i}$ acts on the system as

$$\boldsymbol{\Lambda}_{int_i} = \boldsymbol{J}_{H_i} \boldsymbol{\Lambda}^B_{int_i} = \left[\begin{bmatrix} \boldsymbol{I}_3 & -[\boldsymbol{R}^H_B \boldsymbol{p}_i]_\wedge \\ \boldsymbol{0}_3 & \boldsymbol{I}_3 \end{bmatrix} \begin{bmatrix} \boldsymbol{R}^H_B & \boldsymbol{0}_3 \\ \boldsymbol{0}_3 & \boldsymbol{I}_3 \end{bmatrix}\right]^T \boldsymbol{\Lambda}^B_{int_i}, \tag{3.3}$$

where $\boldsymbol{0}_3$ and $\boldsymbol{I}_3$ are respectively the 3×3 null and identity matrices.

Denoting with $\boldsymbol{\Lambda}_{int} = [\boldsymbol{F}^T_{int}\ \boldsymbol{\tau}^T_{int}]^T \in \mathbb{R}^6$ the resultant of the interaction wrenches, then the total external wrench is

$$\boldsymbol{\Lambda}_{ext_{ex}} = \boldsymbol{\Lambda}_{dis} + \boldsymbol{\Lambda}_{int} = \boldsymbol{J}_{H_{dis}} \boldsymbol{\Lambda}^B_{dis} + \sum_{i=1}^{q} \boldsymbol{J}_{H_i} \boldsymbol{\Lambda}^B_{int_i}. \tag{3.4}$$

3.2.3 Force/Torque detectors

The UAV is equipped with n sensing devices $s_1, \dots, s_n$ attached to n points $\boldsymbol{p}_{s_i}, i = 1, \dots, n$ (in $\mathcal{F}_B$) that are able to measure a force and/or torque applied in their particular location. In the following, we will refer to the $\boldsymbol{p}_{s_i}$'s as Points of Contact (PoC). The measured forces/torques are provided as m-bit quantized signals, with a quantization interval for s_i of $\boldsymbol{F}_{sa_i}$ for the forces (if measured) and $\boldsymbol{\tau}_{sa_i}$ for the torques (if measured).

This generic formulation includes different types of sensor devices, from proper force/-torque sensors to simple push buttons. In our hardware setup we will use the latter, hence we proceed here to particularize their case. However, most of the findings in this work are compatible with more complex sensing devices.

Push buttons can be modeled as simple 1-bit quantization force sensors measuring a force acting along the normal vector to the button surface. Let be s_i a push button sensor, and let be

$$\boldsymbol{n}_i = [\cos(\alpha_i)\ \cos(\beta_i)\ \cos(\gamma_i)]^T \tag{3.5}$$

the incoming normal vector to its surface in $\mathcal{F}_B$, specified through its direction cosines. In the previous expression, α_i, β_i, γ_i are the angles between $\boldsymbol{n}_i$ and the axes $\boldsymbol{X}_B, \boldsymbol{Y}_B, \boldsymbol{Z}_B$ respectively.

Then, the sensor can provide two possible measurement: (i) 0, which means that there is no detected force applied in $\boldsymbol{p}_{s_i}$, hence s_i is inactive, and (ii) 1, which means that there is a force $\boldsymbol{F}^B_{int_i} = \|\boldsymbol{F}^B_{int_i}\|\boldsymbol{n}_i$ applied in $\boldsymbol{p}_{s_i}$, with $\|\boldsymbol{F}^B_{int_i}\| > \boldsymbol{F}_{sa_i}$, and in this case s_i is active.

In the rest of the chapter, we assume that at a given time instant only $N \leq n$ out of $\{s_1,\ldots,s_n\}$ are active. Without loss of generality we rename the active sensors, their PoCs, their quantization interval and the normal vector to their surface respectively S_i, $\boldsymbol{P}_{S_i}$, $\boldsymbol{F}_{Sa_i}$ and $\boldsymbol{N}_i$, for $i = 1,\ldots,N$. We will also denote with $\boldsymbol{\Lambda}^B_{int_i} = [\boldsymbol{F}^{B^T}_{int_i}\ 0\ 0\ 0]^T$ the interaction wrench acting on $\boldsymbol{P}_{S_i}$.

3.3 System Architecture

The main goal of this work is to develop an estimation and control framework such that physical interactions of humans with the UAV results in actions from the quadrotor that second the interaction forces.

Our system architecture (see Fig. 3.3) is composed of two main block chains. The first estimation chain is in charge of estimating the relevant dynamic quantities that are needed in the second control chain. In particular, the system will compute estimates $\hat{\boldsymbol{\Lambda}}^B_{int_i}, i = 1,\ldots,N$, $\hat{\boldsymbol{\Lambda}}^B_{dis}$ of the interaction and disturbance wrenches respectively, based on the knowledge of ρ, $\boldsymbol{\tau}$, $\boldsymbol{\xi}_H$, $\boldsymbol{\zeta}$, S_i, $\boldsymbol{P}_{S_i}$, $\boldsymbol{F}_{Sa_i}$ and $\boldsymbol{N}_i$, for $i = 1,\ldots,N$.

The estimation chain consists of a two step system. In the first step, the system state $\dot{\boldsymbol{p}}_H$, as well as the control commands ρ, $\boldsymbol{\tau}$ are used to compute a minimal error estimate $\hat{\boldsymbol{\Lambda}}_{ext}$ of the total external wrench $\boldsymbol{\Lambda}_{ext}$ acting on the quadrotor, using a momenta based residual estimator.

This estimate is then decomposed into estimates of the interaction $\hat{\boldsymbol{\Lambda}}^B_{int_i}, i = 1,\ldots,N$ and disturbance wrenches $\hat{\boldsymbol{\Lambda}}^B_{dis}$ according to equation (3.4), in which interaction forces are considered to be applied on the active sensors $S_1,\ldots,S_N$. In this step, the sensor readings, as well as the sensor parameters are used into a quadratic programming problem which tries to explain the estimated $\hat{\boldsymbol{\Lambda}}_{ext}$ using the minimum norm stacked vector $[\hat{\boldsymbol{\Lambda}}^{B\,T}_{dis}\ \hat{\boldsymbol{\Lambda}}^{B\,T}_{int_1}\ \ldots\ \hat{\boldsymbol{\Lambda}}^{B\,T}_{int_N}]^T$.

The control chain is designed to perform a trajectory tracking. In the beginning, the desired trajectory $\boldsymbol{p}_d$, $\dot{\boldsymbol{p}}_d$, $\ddot{\boldsymbol{p}}_d$ is produced by a trajectory generator. However, when the interaction wrench estimates are nonzero, an admittance controller modifies it based on their resultant $\hat{\boldsymbol{\Lambda}}_{int}$. The new reference trajectory is passed to a geometric trajectory tracking controller which uses the knowledge of the estimated disturbance wrench $\hat{\boldsymbol{\Lambda}}_{dis}$ to reject it through a feedforward term.

3.4 Hardware-Software Design

The UAV platform employed in this work is also MK-Quadro quadrotor as utilized earlier in the wrench estimation in Chapter. 2. It consists of propeller arms , motor controllers, brushless motors and 10 inch propellers. Furthermore it consists of microcon-

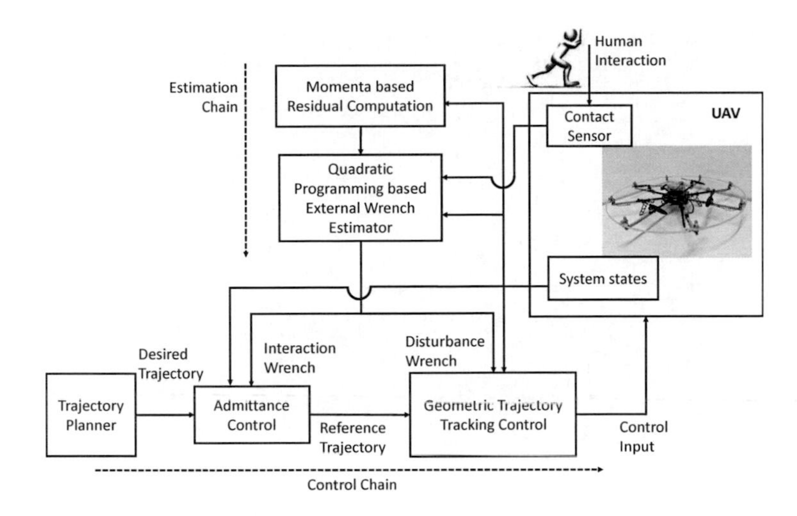

Figure 3.3: Block scheme of the system architecture for human-UAV interaction.

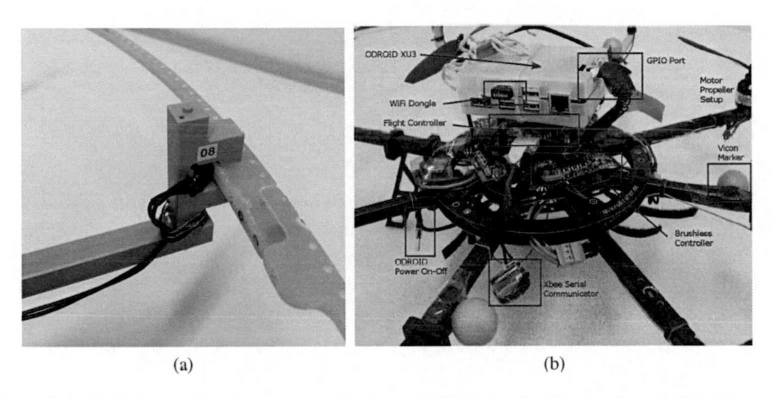

(a) (b)

Figure 3.4: Experimental hardware setup for human-UAV physical interaction. 3.4(a) Close-up view of one of the arm extenders with the button and the interaction ring. 3.4(b) Zoomed in figure showing the Odroid XU3 board, GPIO access port, Wifi communicator, flight controller, brushless controller, Xbee serial communicator, motor-propeller setup and vicon markers.

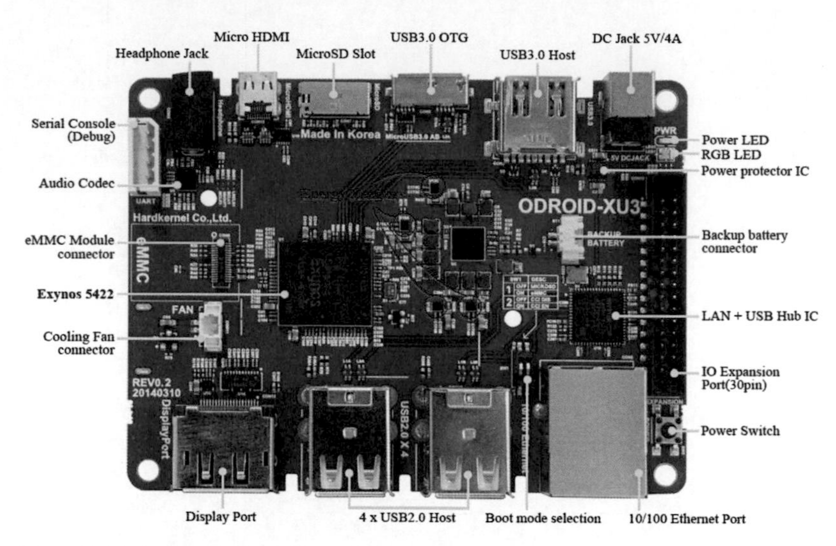

Figure 3.5: ODROID XU3 used in the hardware setup. The 30 pin GPIO Expansion Port was utilized to obtain the point of contact sensor state. *Source: www.hardkernel.com.*

troller, which is used to perform the low-level control, transferring the control commands from the high-level controller to the motor controller. The microcontroller board includes an inertial measurement unit (IMU) composed of an accelerometer and a gyroscope. The command sent to the microcontroller at $\sim$120 Hz, consist of the setpoints for the brushless motors, which are computed on an offboard desktop PC. The platform is powered by a 2600 mAh LiPo battery that provides approximately 10 min of flight. The complete details of the hardware components were detailed earlier in Sec. 2.6. The hardware setup used is shown in Fig. 3.4(b). Additionally, the quadrotor is equipped with Odroid-XU3 (shown in Fig. 3.5), a double quad core ARM microprocessor board. The power to the Odroid and its components are provided by a $5\,V$ step-down voltage regulator connected to the LiPo battery. The Odroid is also fitted with Wi-Fi adapter which is useful for exchanging data with the ground station.

The UAV, whose picture is shown in Fig. 3.1, is retrofitted with four additional arms. Extenders at the end of each arm are fixed so that the whole propeller is always inside the diameter of the UAV setup. One 12 mm square momentary button is mounted at the end of each extender. A floating ring structure encloses the whole quadrotor while being always in contact with the buttons as shown in Fig. 3.4(a). In this configuration, the buttons are all coplanar and lie on the same XY-plane of the CoM. Moreover, they are equally spaced along the inner circumference of the ring with an angular distance of 45°.

In this way, a force applied to any point of the ring in the direction of the CoM of the quadrotor activates one or more buttons depending on the location of the point of contact. For example, if the PoC is near one button, that button will be activated, whereas a PoC between two buttons, will cause the activation of both. Naming the sensors as $s_1, \ldots, s_8$ (as shown in Fig. 3.2) the interaction points and the normal vectors are in the form

$$\boldsymbol{p}_{s_i} = \boldsymbol{R}_z\left(\frac{\pi(i-1)}{4}\right) L e_1, \quad \boldsymbol{n}_i = -\boldsymbol{R}_z\left(\frac{\pi(i-1)}{4}\right) e_1,$$

where $L = 0.34\,m$ is the length of the arms and $e_1 = [1\;0\;0]^T$. The buttons are connected to the General-Purpose Input/Output (GPIO) ports of the Odroid-XU3 board which communicates them to the base station through a Robot Operating System (ROS) topic over Wi-Fi connection as shown in Fig. 3.4(b).

The main control and estimation algorithms are performed in the base station, which is a ROS enabled Ubuntu 14.04 PC. The state of the UAV in the $10m \times 10m$ flying arena is provided at 120 Hz by a motion capture system (VICON), which is also used to collect ground truth data.

3.5 Estimation of the External, Interactive and Disturbance Wrenches

Our goal is to estimate the interaction $\boldsymbol{\Lambda}^B_{int_i}, i = 1, \ldots, N$ and disturbance $\boldsymbol{\Lambda}^B_{dis}$ wrenches acting on the UAV. In order to achieve this, we follow a two-step procedure. First, we employ a residual based estimator to estimate the total external wrench $\boldsymbol{\Lambda}_{ext_{ex}}$ which comprises of both the interaction and disturbance wrenches as modeled in (3.4). Then, we decompose this wrench in multiple interaction components $\boldsymbol{\Lambda}^B_{int_i}, i = 1, \ldots, N$ and disturbance component $\boldsymbol{\Lambda}^B_{dis}$, based on a least square principle after defining a suitable optimization problem.

3.5.1 Estimation of the External Wrench

The estimation of $\boldsymbol{\Lambda}_{ext_{ex}}$ of extended wrench model is similar to Sec. 2.3 where, utilizing the Lagrangian quadrotor model, the residual based procedure on the idea of generalized momenta gives the residual vector given by

$$\boldsymbol{r} \simeq \boldsymbol{\Lambda}_{ext_{ex}}, \tag{3.6}$$

which in turn is used to obtain $\hat{\boldsymbol{\Lambda}}_{ext_{ex}}$ given by

$$\hat{\boldsymbol{\Lambda}}_{ext_{ex}} = \boldsymbol{r}. \tag{3.7}$$

Refer Sec. 2.3 for the detailed summary of the approach used to obtain estimation of extended external wrench (3.7) because the same principle has been followed in this case as well. The objective question of separating the disturbance $\mathbf{\Lambda}_{dis}$ and the interaction $\mathbf{\Lambda}_{int}$ wrenches is therefore the main focus in Sec. 3.5.2.

3.5.2 Estimation of the disturbance and interaction wrenches

Clearly, the estimate of the external wrench from (3.7) accounts for both the interaction and disturbance wrenches. The interaction wrench further consists of the wrenches arising from different PoC, as specified in equation (3.4). In order to allow a safe UAV-HRPI, we need to compute separate estimates for all those components.

Here, we have considered that the wrenches arising from the human interaction are observed at one or more of the preselected points $\boldsymbol{p}_{s_i}$, where the sensors are installed. This assumption substantially reduces the number of constraints in the quadratic programming based optimization giving it a computational edge over considering a sensor surface (e.g. tactile sensor setup). However, provided that the sensors are deployed so that they can correctly describe any possible interaction wrench, this advantage comes at no price, since the admittance controller will use the resultant of the wrenches estimated at the sensors' location (see Sec. 3.6.1) and not their specific values. Conversely, if the sensors are not located correctly a fraction or the whole interaction force may be interpreted as disturbance.

By stacking the transposed matrices $\boldsymbol{J}_{H_{dis}}$, $\boldsymbol{J}_{H_i}$ and the disturbance and interaction wrenches $\mathbf{\Lambda}^B_{dis}$, $\mathbf{\Lambda}^B_{int_i}$, the external wrench $\mathbf{\Lambda}_{ext_{ex}}$ from equation (3.4) can be written as

$$\mathbf{\Lambda}_{ext_{ex}} = \begin{bmatrix} \boldsymbol{J}_{H_{dis}} & \boldsymbol{J}_{H_1} & \cdots & \boldsymbol{J}_{H_N} \end{bmatrix} \begin{bmatrix} \mathbf{\Lambda}^B_{dis} \\ \mathbf{\Lambda}^B_{int_1} \\ \vdots \\ \mathbf{\Lambda}^B_{int_N} \end{bmatrix}. \tag{3.8}$$

In the above expression, all $\mathbf{\Lambda}^B_{int_i}$ are applied in their real point of contact, whereas $\mathbf{\Lambda}^B_{dis}$ is acting in the center of mass of the quadrotor. Being $\hat{\mathbf{\Lambda}}^B_{dis}$, $\hat{\mathbf{\Lambda}}^B_{int_i}$ the estimates of Λ^B_{dis}, $\Lambda^B_{int_i}$, they can be related to the residual vector from equation (3.7):

$$\boldsymbol{r} = \hat{\mathbf{\Lambda}}_{ext_{ex}} = \underbrace{\begin{bmatrix} \boldsymbol{J}_{H_{dis}} & \boldsymbol{J}_{H_1} & \cdots & \boldsymbol{J}_{H_N} \end{bmatrix}}_{\triangleq \boldsymbol{A}_1} \underbrace{\begin{bmatrix} \hat{\mathbf{\Lambda}}^B_{dis} \\ \hat{\mathbf{\Lambda}}^B_{int_1} \\ \vdots \\ \hat{\mathbf{\Lambda}}^B_{int_N} \end{bmatrix}}_{\triangleq \hat{\mathbf{\Lambda}}^B_c}, \tag{3.9}$$

where the matrix $\boldsymbol{A}_1 \in \mathbb{R}^{6\times 6(N+1)}$ has always $\text{rank}(\boldsymbol{A}_1) = 6$.

In general, for $N > 0$ (hence when at least one interaction force is present) the problem of finding $\hat{\boldsymbol{\Lambda}}_c^B$ from $\hat{\boldsymbol{\Lambda}}_{ext_{ex}}$ by inverting equation (3.9) always admits infinite solutions. Therefore, we formulate the estimation of $\hat{\boldsymbol{\Lambda}}_c^B$ as the solution of a quadratic optimization problem in the form:

$$\hat{\boldsymbol{\Lambda}}_c^B = \underset{\boldsymbol{\lambda}\in\mathbb{R}^p}{\text{argmin}}\, \boldsymbol{\lambda}^T \boldsymbol{D}\boldsymbol{\lambda} + 2\boldsymbol{C}^T\boldsymbol{\lambda} \tag{3.10}$$

$$\text{s.t. } \boldsymbol{A}\boldsymbol{\lambda} \leq \boldsymbol{b} \tag{3.11}$$

where $p = 6(N+1)$, $\boldsymbol{D} \in \mathbb{R}^{p\times p}$ and $\boldsymbol{C} \in \mathbb{R}^{p\times 6}$ are appropriate weight matrices and $\boldsymbol{A} \in \mathbb{R}^{p\times q}$ and $\boldsymbol{b} \in \mathbb{R}^q$ defines q appropriate constraints.

Objective function. In the choice of $\boldsymbol{D}$ and $\boldsymbol{C}$, we want to follow two main principles. First, since infinite solutions are possible, we want to have the minimal wrench configuration that explains the readings of the sensors and the estimated external wrench. Therefore we choose $\boldsymbol{C} = \boldsymbol{0}_{p\times 6}$ and we select $\boldsymbol{D}$ as a positive definite diagonal matrix whose eigenvalues are strictly positive. This choice ensures that there is a unique global minimum while the quadratic problem is solved. Secondly, the disturbance wrench $\hat{\boldsymbol{\Lambda}}_{dis}^B$ should contain only those component of $\hat{\boldsymbol{\Lambda}}_{ext_{ex}}$ that cannot be explained through interaction wrenches according to the sensor readings. Hence, we penalize $\hat{\boldsymbol{\Lambda}}_{dis}^B$ through a matrix in the form

$$\boldsymbol{D} = \begin{pmatrix} w\boldsymbol{I}_6 & \boldsymbol{0}_6 & \cdots & \boldsymbol{0}_6 \\ \boldsymbol{0}_6 & \boldsymbol{I}_6 & \cdots & \boldsymbol{0}_6 \\ \vdots & \vdots & \ddots & \vdots \\ \boldsymbol{0}_6 & \boldsymbol{0}_6 & \cdots & \boldsymbol{I}_6 \end{pmatrix}, \tag{3.12}$$

where $w = 50 \gg 1$ is an appropriate weight. One limitation of this approach arises when both the disturbance and the interaction wrenches are in the same direction, since the disturbance can be explained by the sensor readings and will be accounted as additional interaction wrench. In order to mitigate such issue, it is possible to study an adaptive law which varies w over time. Note that the effect of a disturbance in this scenario will affect mainly the absolute value of the estimated interaction force, while it will not affect its direction.

In the following, we consider four constraints. While the first and the fourth are valid in general, the other two are specific for the type of sensor that we have used. Other constraints can be considered if the UAV is equipped with other types of sensors.

Constraint 1. *$\hat{\boldsymbol{\Lambda}}_c^B$ must respect $\boldsymbol{A}_1\hat{\boldsymbol{\Lambda}}_c^B = \hat{\boldsymbol{\Lambda}}_{ext_{ex}} = \boldsymbol{r}$.*

This constraint is the direct application of equation (3.9).

Constraint 2. *The interaction force $\boldsymbol{F}_{int_i}^B$ lies along the normal vector $\boldsymbol{N}_i$.*

The equation for this constraint can be found by taking into account the direction cosines of $\boldsymbol{N}_i$ from equation (3.5). The relation between the direction cosines and the interaction forces for sensor S_i is

$$\left|\hat{\boldsymbol{F}}^B_{int_i}\right| = \frac{\hat{\boldsymbol{F}}^B_{intx_i}}{\cos(\alpha_i)} = \frac{\hat{\boldsymbol{F}}^B_{inty_i}}{\cos(\beta_i)} = \frac{\hat{\boldsymbol{F}}^B_{intz_i}}{\cos(\gamma_i)}. \tag{3.13}$$

The following three constraints are equivalent to equation (3.13)

$$\begin{aligned} \hat{\boldsymbol{F}}^B_{intx_i}\cos(\beta_i) - \hat{\boldsymbol{F}}^B_{inty_i}\cos(\alpha_i) = 0 \\ \hat{\boldsymbol{F}}^B_{intx_i}\cos(\gamma_i) - \hat{\boldsymbol{F}}^B_{intz_i}\cos(\alpha_i) = 0 \\ \hat{\boldsymbol{F}}^B_{inty_i}\cos(\gamma_i) - \hat{\boldsymbol{F}}^B_{intz_i}\cos(\beta_i) = 0 \end{aligned} \tag{3.14}$$

which in matrix form becomes

$$\underbrace{\begin{bmatrix} \cos(\beta_i) & -\cos(\alpha_i) & 0 \\ \cos(\gamma_i) & 0 & -\cos(\alpha_i) \\ 0 & \cos(\gamma_i) & -\cos(\beta_i) \end{bmatrix}}_{\triangleq \boldsymbol{A}_{2_i}} \begin{bmatrix} \hat{\boldsymbol{F}}^B_{intx_i} \\ \hat{\boldsymbol{F}}^B_{inty_i} \\ \hat{\boldsymbol{F}}^B_{intz_i} \end{bmatrix} = \underbrace{\begin{bmatrix} 0 \\ 0 \\ 0 \end{bmatrix}}_{\triangleq \boldsymbol{b}_{2_i}}. \tag{3.15}$$

Note that the constraints in (3.15) exists for each active sensor $S_1, \ldots, S_N$.

Constraint 3. *Each sensor has a minimum activation force* $\boldsymbol{F}_{Sa_i} \in \mathbb{R}$.

Let $\boldsymbol{F}_{Sa_i}$ be the minimum force required by the i-th sensor to be activated. Then

$$\left|\hat{\boldsymbol{F}}^B_{int_i}\right| \geq \boldsymbol{F}_{Sa_i}. \tag{3.16}$$

By using the expression of the direction cosines to describe the force applied to each S_i, we obtain three more constraints for each sensor, which in matrix form are

$$\begin{bmatrix} \hat{\boldsymbol{F}}^B_{intx_i} \\ \hat{\boldsymbol{F}}^B_{inty_i} \\ \hat{\boldsymbol{F}}^B_{intz_i} \end{bmatrix} \geq \underbrace{\begin{bmatrix} \boldsymbol{F}_{Sa_i}\cos(\alpha_i) \\ \boldsymbol{F}_{Sa_i}\cos(\beta_i) \\ \boldsymbol{F}_{Sa_i}\cos(\gamma_i) \end{bmatrix}}_{\triangleq \boldsymbol{b}_{3_i}}. \tag{3.17}$$

Constraint 4. *The interaction torques* $\boldsymbol{\tau}_{intx_i}$, $\boldsymbol{\tau}_{inty_i}$ *are always zero.*

This constraint is necessary because the torques around $\boldsymbol{X}_B$ and $\boldsymbol{Y}_B$ directly affect the rotational dynamics of the quadrotor (2.6). Since the quadrotor is an underactuated system, the roll and pitch angles cannot be selected independently.

However, if the UAV is fully actuated (refer Chapter. 5 (Rajappa *et al.*, 2015)) this constraint can be reformulated to allow admittance of the interaction torques around $\boldsymbol{X}_B$

and $\boldsymbol{Y}_B$. In matrix form, this constraint is

$$\begin{bmatrix} \hat{\boldsymbol{\tau}}_{int_{x_i}} \\ \hat{\boldsymbol{\tau}}_{int_{y_i}} \end{bmatrix} = \begin{bmatrix} 0 \\ 0 \end{bmatrix}. \tag{3.18}$$

Remark 3. *The interaction torque $\hat{\boldsymbol{\tau}}_{int_{z_i}}$ is not constrained to be identically null and it is estimated alongside $\hat{\boldsymbol{F}}^B_{int_i}$ because yaw can be set independently and does not get affected by the underactuation problem.*

Optimization Problem. The final form of the quadratic optimization problem is

$$\hat{\boldsymbol{\Lambda}}^B_c = \underset{\boldsymbol{\lambda} \in \mathbb{R}^p}{\operatorname{argmin}} \, \boldsymbol{\lambda}^T \boldsymbol{D} \boldsymbol{\lambda} \tag{3.19}$$
$$\text{s.t.} \begin{cases} \boldsymbol{A}_1 \boldsymbol{\lambda} = \boldsymbol{r} \\ \boldsymbol{A}_{2_i} \boldsymbol{\lambda}_i = \boldsymbol{b}_{2_i}, & i = 1, \ldots, N \\ \boldsymbol{\lambda}_i \geq \boldsymbol{b}_{3_i}, & i = 1, \ldots, N \\ \boldsymbol{\lambda}_{\boldsymbol{\tau}_i} = 0, & i = 1, \ldots, N \end{cases}$$

where $\boldsymbol{\lambda}_i \in \mathbb{R}^3$ is the vector of the variables in $\boldsymbol{\lambda}$ corresponding to $\hat{\boldsymbol{F}}^B_{int_i}$, $\boldsymbol{\lambda}_{\boldsymbol{\tau}_i} \in \mathbb{R}^2$ is the vector of the variables in $\boldsymbol{\lambda}$ corresponding to $[\hat{\boldsymbol{\tau}}_{int_{x_i}} \; \hat{\boldsymbol{\tau}}_{int_{y_i}}]^T$, $\boldsymbol{A}_1$ is defined in (3.9), $\boldsymbol{r}$ is the residual computed in (3.6), $\boldsymbol{A}_{2_i}$ along with $\boldsymbol{b}_{2_i}$ are defined in (3.15) and $\boldsymbol{b}_{3_i}$ is defined from (3.17).

In our experimental setup the problem defined in equation (3.19) is solved at every time step through the quadratic solver included in the Computational Geometry Algorithms Library (CGAL)[1].

3.6 Control

In this section, we design a control scheme to drive the UAV based on the external force and torques. The main goal of the control framework is to admit the estimated interaction wrench $\hat{\boldsymbol{\Lambda}}_{int}$ while rejecting the disturbance wrench $\hat{\boldsymbol{\Lambda}}_{dis}$. Therefore, these two components will be treated differently inside the controller, which can be divided into two main parts: (i) the high level admittance control scheme that uses $\hat{\boldsymbol{\Lambda}}_{int}$ to compute a reference trajectory and (ii) the low level trajectory tracking which is in charge of rejecting $\hat{\boldsymbol{\Lambda}}_{dis}$.

3.6.1 Admittance Control

In the admittance control framework the desired trajectory $\boldsymbol{p}_d(t)$, $\dot{\boldsymbol{p}}_d(t)$, $\ddot{\boldsymbol{p}}_d(t)$ in $\mathcal{F}_W$ is modified based on the estimated interaction forces $\hat{\boldsymbol{F}}^B_{int_i}, i = 1, \ldots, N$ to provide a reference trajectory for the low level controller $\boldsymbol{p}_a(t)$, $\dot{\boldsymbol{p}}_a(t)$, $\ddot{\boldsymbol{p}}_a(t)$. Let the admittance force

[1] http://www.cgal.org/

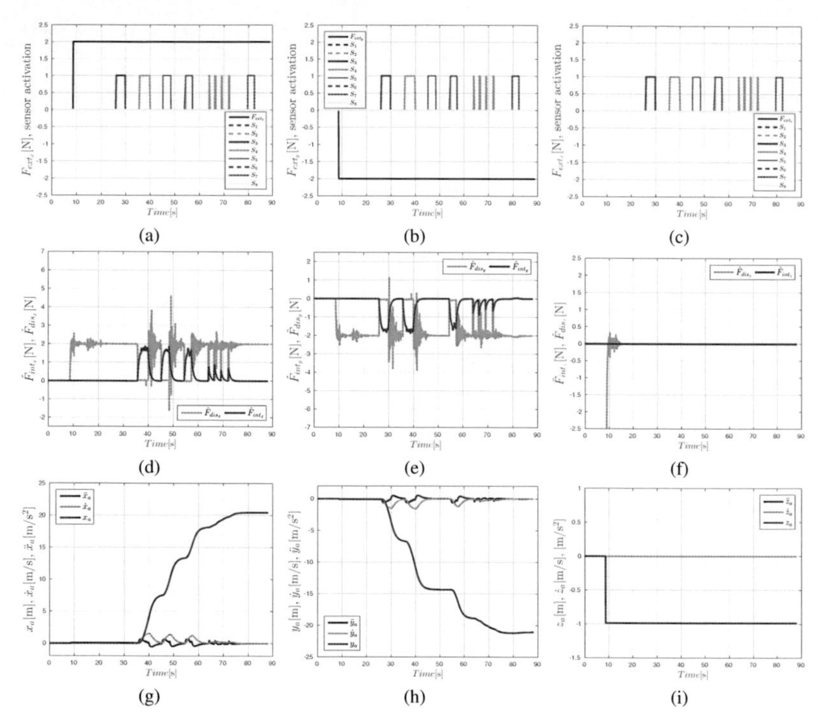

Figure 3.6: Results of hardware-in-the-loop simulations. 3.6(a): External force applied along X-axis (black) and sensor($s_i, \forall\, i = 1 \rightarrow 8$) activation status for s_1 (dashed red), s_2 (dashed green), s_3 (solid red), s_4 (solid green), s_5 (solid blue), s_6 (dashed blue), s_7 (solid magenta), s_8 (solid yellow). 3.6(b): External force applied along Y-axis (black) and sensor($s_i, \forall\, i = 1 \rightarrow 8$) activation status. 3.6(c): External force applied along Z-axis (black) and sensor($s_i, \forall\, i = 1 \rightarrow 8$) activation status. 3.6(d): Estimated interaction force $\hat{\boldsymbol{F}}_{int_x}$ (red) and disturbance force $\hat{\boldsymbol{F}}_{dis_x}$ (green) along X-axis. 3.6(e): Estimated interaction force $\hat{\boldsymbol{F}}_{int_y}$ (red) and disturbance force $\hat{\boldsymbol{F}}_{dis_y}$ (green) along Y-axis. 3.6(f): Estimated interaction force $\hat{\boldsymbol{F}}_{int_z}$ (red) and disturbance force $\hat{\boldsymbol{F}}_{dis_z}$ (green) along Z-axis. 3.6(g): Admittance acceleration $\ddot{x}_a$ (red), admittance velocity $\dot{x}_a$ (green) and admittance position x_a (blue) along X-axis. 3.6(h): Admittance acceleration $\ddot{y}_a$ (red), admittance velocity $\dot{y}_a$ (green) and admittance position y_a (blue) along Y-axis. 3.6(i): Admittance acceleration $\ddot{z}_a$ (red), admittance velocity $\dot{z}_a$ (green) and admittance position z_a (blue) along Z-axis.

$\boldsymbol{F}_a$ expressed in $\mathcal{F}_W$ be defined as the low-pass filtered resultant of all the $\hat{\boldsymbol{F}}_{int_i}$:

$$\boldsymbol{F}_a = \underset{\text{pass}}{\text{low}}\left(\hat{\boldsymbol{F}}^W_{int}\right) = \underset{\text{pass}}{\text{low}}\left(\sum_{i=1}^{n} \boldsymbol{R}^W_B \hat{\boldsymbol{F}}^B_{int_i}\right). \tag{3.20}$$

In theory, the most faithful behavior would be to use directly $\hat{\boldsymbol{F}}^W_{int}$. However, $\hat{\boldsymbol{F}}^W_{int}$ is a discontinuous signal because it depends on the discretized signal coming from the contact sensors. Therefore, it would cause discontinuous accelerations and commanded attitudes, which are not suitable for safe UAV-HRPI. Hence, the role of the low-pass filter is to smoothen these discontinuities. Here, a 2nd-order low pass filter with filter time constant $t_{fi} = 0.35\ s$ is used. The average time step of the estimator is around $t_{step} = 8\ ms$

In order to modify the desired trajectory, we consider the UAV as an ideal mass-spring-damper system driven by the state equation

$$\ddot{\boldsymbol{p}}_a = \frac{\boldsymbol{F}_a + D(\dot{\boldsymbol{p}}_d - \dot{\boldsymbol{p}}_a) + S(\boldsymbol{p}_d - \boldsymbol{p}_a) + M\ddot{\boldsymbol{p}}_d}{M}, \tag{3.21}$$

where $M \in \mathbb{R}^+$ is the virtual mass, the diagonal positive semidefinite constant matrices $D, S \in \mathbb{R}^{3\times 3}$ that define a Hurwitz polynomial are the damping and stiffness constants that are used to change the physical properties of the UAV. Note that the elements of D and S are ≥ 0. The values M, D and S can be chosen in order to provide a human friendly behavior avoiding sudden accelerations and allowing to exert forces on the UAV. In general, their value can be selected independently on each axis. However, in our case we have chosen $D = d\boldsymbol{I}_3$ with $d = 1$ and $S = s\boldsymbol{I}_3$, with $s = 0$ during human friendly interaction in Sec. 3.8.1 or varying values of s as mentioned in Sec. 3.8.2 for the trajectory tracking. In order to have a complete reference trajectory in the form $\boldsymbol{p}_a(t)$, $\dot{\boldsymbol{p}}_a(t)$, $\ddot{\boldsymbol{p}}_a(t)$, the values of $\dot{\boldsymbol{p}}_a$ and $\boldsymbol{p}_a$ are computed by integrating $\ddot{\boldsymbol{p}}_a$ in time.

3.6.2 Trajectory Tracking Control with Wrench Feedforward

In order to command the UAV to follow the reference trajectory, we use a control law based on the one proposed by Lee *et al.* (2010) and improved by Spica *et al.* (2013) because of its global convergence, aggressive maneuvers capability and excellent trajectory tracking performance. In addition, the controller for the rotational dynamics is developed directly on $SO(3)$ and thereby it avoids any singularities that arise in local coordinates, such as Euler angles. In order to reject the estimated disturbance wrench, we include a feedforward disturbance compensation term.

Considering the trajectory tracking task, at a given time step the tracking error in position and velocity are defined as $\boldsymbol{e}_p = \boldsymbol{p}_W - \boldsymbol{p}_a$ and $\boldsymbol{e}_v = \dot{\boldsymbol{p}}_W - \dot{\boldsymbol{p}}_a$ respectively. The desired force for the translational dynamics is given as,

$$\rho = (m\ddot{\boldsymbol{p}}_a - \boldsymbol{K}_d\boldsymbol{e}_v - \boldsymbol{K}_p\boldsymbol{e}_p - \\ - \boldsymbol{K}_i \int_{t_0}^{t} \boldsymbol{e}_p dt - mge_3 - \hat{\boldsymbol{F}}_{dis}) \cdot \boldsymbol{R}_B^W e_3, \tag{3.22}$$

where the diagonal positive definite gain matrices $\boldsymbol{K}_d$, $\boldsymbol{K}_p$, $\boldsymbol{K}_i$ define Hurwitz polynomials. The desired hovering thrust is realized by $f_z = \rho\, e_3$ and by aligning the body vertical axis along the direction of the ρ defined as,

$$\vec{z}_{R_d} = \frac{m\ddot{\boldsymbol{p}}_a - \boldsymbol{K}_d\boldsymbol{e}_v - \boldsymbol{K}_p\boldsymbol{e}_p - \boldsymbol{K}_i \int_{t_0}^{t} \boldsymbol{e}_p dt - mge_3 - \hat{\boldsymbol{F}}_{dis}}{\|m\ddot{\boldsymbol{p}}_a - \boldsymbol{K}_d\boldsymbol{e}_v - \boldsymbol{K}_p\boldsymbol{e}_p - \boldsymbol{K}_i \int_{t_0}^{t} \boldsymbol{e}_p dt - mge_3 - \hat{\boldsymbol{F}}_{dis}\|}, \tag{3.23}$$

where $\vec{z}_{R_d}$ is the third column of the desired attitude rotation matrix $\boldsymbol{R}_{B_d}^W$ defined as $\boldsymbol{R}_{B_d}^W = [\vec{x}_{R_d}, \vec{y}_{R_d}, \vec{z}_{R_d}] \in SO(3)$. Since the quadrotor UAV is an underactuated system, the desired attitude generated by the outer-loop translational dynamics is controlled by means of the inner-loop torques, that are generated for controlling the rotational dynamics, to track a desired attitude rotation $\boldsymbol{R}_{B_d}^W$. The other two columns $\vec{x}_{R_d}$ and $\vec{y}_{R_d}$ of $\boldsymbol{R}_{B_d}^W$, which account for the remaining degrees of freedom, should be chosen such that their direction is orthogonal to $\vec{z}_{R_d}$ and minimize the yaw error. Therefore

$$\vec{x}_{R_d} = \vec{y}_{R_d} \times \vec{z}_{R_d}, \qquad \vec{y}_{R_d} = \frac{\vec{z}_{R_d} \times \vec{x}_{R_d}}{\|\vec{z}_{R_d} \times \vec{x}_{R_d}\|}. \tag{3.24}$$

For the rotational dynamics, assuming that $\boldsymbol{\omega}_{B_d} = [\boldsymbol{R}_{B_d}^{W\,T} \dot{\boldsymbol{R}}_{B_d}^{W\,T}]_\vee$, where $[\cdot]_\vee$ represents the inverse (vee) operator from $so(3) \to \mathbb{R}^3$, the attitude tracking error $\boldsymbol{e}_R \in \mathbb{R}^3$ is defined similarly to Lee *et al.* (2010) as

$$\boldsymbol{e}_R = \frac{1}{2}[\boldsymbol{R}_{B_d}^{W\,T} \boldsymbol{R}_B^W - \boldsymbol{R}_B^{W\,T} \boldsymbol{R}_{B_d}^W]_\vee, \tag{3.25}$$

and the tracking error of the angular velocity $\boldsymbol{e}_\omega \in \mathbb{R}^3$ is given by

$$\boldsymbol{e}_\omega = \boldsymbol{\omega}_B - \boldsymbol{R}_B^{W\,T} \boldsymbol{R}_{B_d}^W \boldsymbol{\omega}_{B_d}. \tag{3.26}$$

In order to obtain an asymptotic convergence to $\mathbf{0}$ of the rotational error $\boldsymbol{e}_R$ one can choose the following controller

$$\boldsymbol{\tau} = -\boldsymbol{K}_\omega \boldsymbol{e}_\omega - \boldsymbol{K}_r \boldsymbol{e}_R - \boldsymbol{K}_{ir} \int_{t_0}^{t} \boldsymbol{e}_R + \boldsymbol{\omega}_B \times \boldsymbol{I}_B \boldsymbol{\omega}_B - \\ - \boldsymbol{I}_B([\boldsymbol{\omega}_B]_\wedge \boldsymbol{R}_B^{W\,T} \boldsymbol{R}_{B_d}^W \boldsymbol{\omega}_{B_d} - \boldsymbol{R}_B^{W\,T} \boldsymbol{R}_{B_d}^W \dot{\boldsymbol{\omega}}_{B_d}) - \hat{\boldsymbol{\tau}}_{dis}, \tag{3.27}$$

where the diagonal positive-definite gain matrices $\boldsymbol{K}_{\omega}$, $\boldsymbol{K}_{r}$, $\boldsymbol{K}_{ir}$ define Hurwitz polynomials, $[\boldsymbol{\omega}_B]_\wedge$ is the skew symmetric matrix of $\boldsymbol{\omega}_B$ and $\hat{\boldsymbol{\tau}}_{dis}$ is the external torque disturbance.

3.7 Hardware-in-the-loop Physical Simulations

In order to verify the wrench estimators and the admittance controller, it is necessary to test the proposed algorithms with the ground truth on the forces and torques applied on the UAV. Recently, it is becoming a trend to perform Hardware-in-the-loop (HIL) simulations to verify various algorithms before their testing on real robot (Chandhrasekaran and Choi, 2010; Cai *et al.*, 2008; Odelga *et al.*, 2015).

In our case, the advantage of performing HIL simulations are (i) the hardware and sensors, the software setup and the communication link with the UAV can be tested as in real experiments; (ii) the wrench estimators can be verified with ground truth data; (iii) the functionality of the admittance control and the disturbance compensator can be verified without exposing people to any danger; and (iv) the computational times and the feasibility in real-time is tested.

Our HIL simulation setup consists of (i) Gazebo[2], a popular open source ROS-enabled simulator, which provides the dynamical simulation of the UAV and feedbacks the corresponding sensor readings (IMU, pose); (ii) our UAV and contact sensor setup, including the ODROID XU3 board which reads the contact sensor state and communicate it to the base station; and (iii) a collection of ROS nodes based on Telekyb software (Grabe *et al.*, 2013) which provide the hardware interfacing, estimation and control functionalities.

Figure. 3.7 shows the block scheme of our HIL simulation setup. The communication between the Odroid board and Gazebo is achieved through ROS topics over IEEE 802.11 connection.

After the takeoff at $t = 8.5\ s$ the HIL physical simulation is carried out by applying for $t \geq 8.5\ s$ an external force of $\boldsymbol{F}_{ext} = \begin{bmatrix}2 & -2 & 0\end{bmatrix}^T N$ in the horizontal frame $\mathcal{F}_H$ while the quadrotor is commanded to stay in hovering. Some PoC sensors s_i are pushed at different time instants between $t = 25\ s$ and $t = 85\ s$. Figure 3.6(a-c) reports the external force and the activation states of the sensors. Figure 3.6(d-f) shows the estimated interaction (red) and disturbance (green) forces throughout the whole simulation. Between $t = 8.5\ s$ and $t = 25\ s$, no PoC sensor is active, hence the whole external force is interpreted as disturbance. The first activated PoC sensors is s_3, which is active between $t = 25\ s$ and $t = 30\ s$. Since s_3 is oriented along the negative Y-axis, during this time the external force along this direction is interpreted as interaction force. Hence, the interaction force estimated is $\hat{\boldsymbol{F}}_{int} \approx \begin{bmatrix}0 & -2 & 0\end{bmatrix}^T N$.

Similarly, interaction forces are estimated as $\hat{\boldsymbol{F}}_{int} = \begin{bmatrix}2 & -2 & 0\end{bmatrix}^T N$ at $t = 35\ s$, $\hat{\boldsymbol{F}}_{int} = \begin{bmatrix}2 & 0 & 0\end{bmatrix}^T N$ at $t = 45\ s$ and $\hat{\boldsymbol{F}}_{int} = \begin{bmatrix}2 & -2 & 0\end{bmatrix}^T N$ at $t = 54\ s$ through the activation of

[2]http://gazebosim.org/

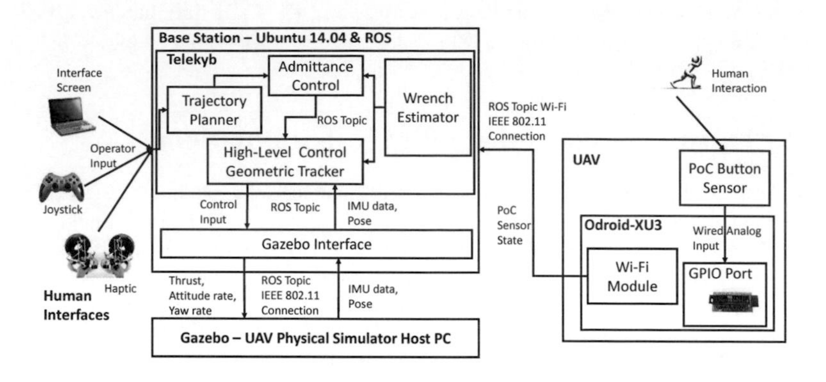

Figure 3.7: Block scheme of hardware-in-the-loop setup

PoC sensors s_4 and s_5. At $t = 54\ s$, a human interaction is made through multiple contact as can be seen that both s_4 and s_5 are active at the same time in Figs. 3.6(a-c). In this case, $\hat{\boldsymbol{F}}_{int} = \boldsymbol{J}_{H_4}\hat{\boldsymbol{F}}^B_{int_4} + \boldsymbol{J}_{H_5}\hat{\boldsymbol{F}}^B_{int_5}$.

Between $t = 60\ s$ and $t = 75\ s$ we tested the scenario in which the UAV is subject to one or more sudden impacts (i.e., the contact happens only for fractions of seconds) as can be seen through the sensor activation of s_3 in Figs. 3.6(a-c). Figures 3.6(d-f) show that during the contact the disturbance falls to zero and the interaction forces rise from zero, hence the UAV is able to detect these impacts. Clearly, the presence of the low pass filter tends to slow down the rate at which the estimated interaction forces rise, so their absolute value is underestimated. This behavior can be mitigated by differently tuning the low pass filter. However, the time constant of the filter used in this work (provided in Sec. 3.6.1) is a good compromise between the need to detect the impacts and the need to provide smooth commanded accelerations and attitude to the UAV.

An interesting case is at $t = 79\ s$. A human interaction happens at s_7, which is in the direction of the positive Y-axis. Although s_7 is active, the external force is $\boldsymbol{F}_{ext} = \begin{bmatrix} 2 & -2 & 0 \end{bmatrix}^T N$, with $2\ N$ force along the negative Y-axis. This indirectly means that there is no interaction force at s_7 despite being active. Hence $\hat{\boldsymbol{F}}_{int}$ in Fig. 3.6(e) detects a small sensor activation force at $t = 79\ s$, whereas all the $\boldsymbol{F}_{ext}$ is considered as disturbance. This once again proves the effectiveness of the estimator during the separation of human interaction forces from the external disturbances.

Figures 3.6(g-i) show the results of the admittance control where the original desired trajectory (hovering) is modified during the human interaction as a new admittance-based trajectory with acceleration $\ddot{\boldsymbol{p}}_a$ (red), velocity $\dot{\boldsymbol{p}}_a$ (green) and position $\boldsymbol{p}_a$ blue, expressed

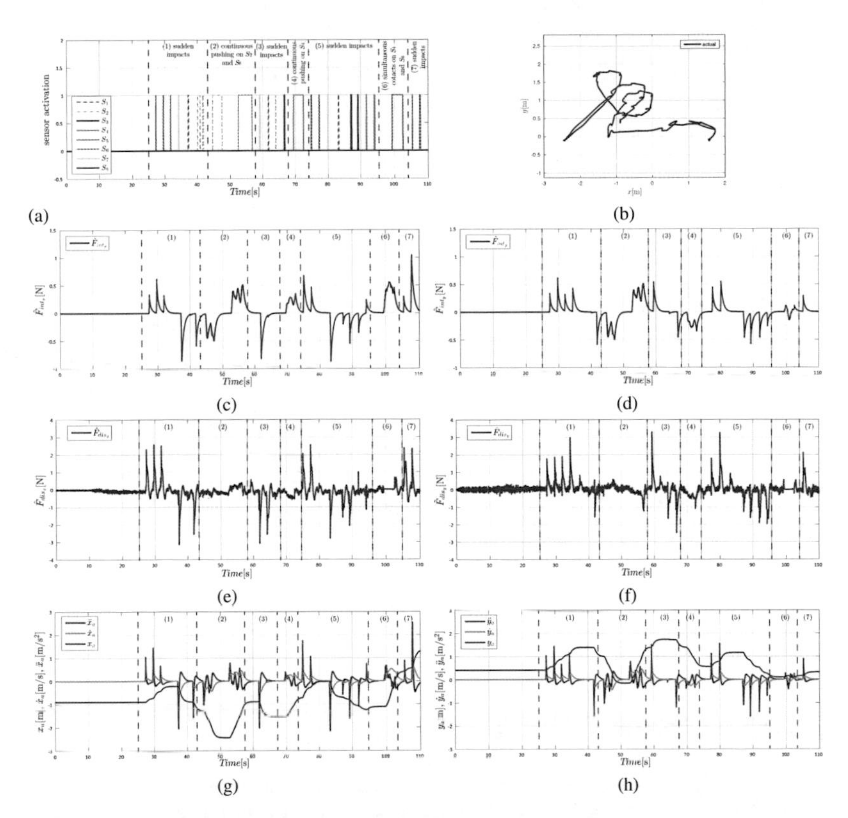

Figure 3.8: Results of continuous pushing, sudden impact and multiple point of contact experiment. 3.8(a): Sensor($s_i, \forall\, i = 1 \rightarrow 8$) activation status for s_1 (dashed red), s_2 (dashed green), s_3 (solid red), s_4 (solid green), s_5 (solid blue), s_6 (solid magenta), s_7 (solid yellow) and s_8 (solid black). 3.8(b): Two dimensional XY-plot of UAV during human interaction. 3.8(c): Estimated interaction force $\hat{\boldsymbol{F}}_{int_x}$ (red) along X-axis. 3.8(d): Estimated interaction force $\hat{\boldsymbol{F}}_{int_y}$ (red) along Y-axis. 3.8(e): Estimated disturbance $\hat{\boldsymbol{F}}_{dis_x}$ (blue) along X-axis. 3.8(f): Estimated disturbance $\hat{\boldsymbol{F}}_{dis_y}$ (blue) along Y-axis. 3.8(g): Admittance acceleration $\ddot{x}_a$ (red), admittance velocity $\dot{x}_a$ (green) and admittance position x_a (blue) along X-axis. 3.8(h): Admittance acceleration $\ddot{y}_a$ (red), admittance velocity $\dot{y}_a$ (green) and admittance position y_a (blue) along Y-axis.

Parameter	Description	Value	Unit
m	mass of the UAV	1.5	Kg
g	gravity acceleration	9.81	m/s^2
I_{xx}	inertia along X-axis	0.011549	$Kg.m^2$
I_{yy}	inertia along Y-axis	0.011368	$Kg.m^2$
I_{zz}	inertia along Z-axis	0.019444	$Kg.m^2$
b	lift coefficient	$1.6073 * 10^{-5}$	N/Ω^2
d	drag coefficient	$2.7988 * 10^{-7}$	Nm/Ω^2
l	arm length	0.315	m
L	distance to sensor	0.34	m
$\boldsymbol{K}_I$	observer gain	5	-
t_{fi}	filter time constant	0.35	s
t_{step}	estimator time step	8	ms
D	damping constant	1	-
S	stiffness constant	0	-
$\boldsymbol{K}_d$	derivative position gain	$diag[5,5,5]$	-
$\boldsymbol{K}_p$	proportional position gain	$diag[25,25,25]$	-
$\boldsymbol{K}_i$	integral position gain	$diag[2,2,2]$	-
$\boldsymbol{K}_\omega$	derivative orientation gain	$diag[0.5,0.5,0.5]$	-
$\boldsymbol{K}_r$	proportional orientation gain	$diag[1.2,1.2,1.2]$	-
$\boldsymbol{K}_{ir}$	integral orientation gain	$diag[0.2,0.2,0.2]$	-

Table 3.1: Experimental parameters

in $\mathcal{F}_H$.

3.8 Experimental Validation

In this section, we present two experiments of Human-UAV interaction performed to test our framework. As the main goal of the previous section was to validate the estimator w.r.t. the ground truth, this section is primarily meant to provide examples of UAV-HRPI and show the feasibility of our approach in real world. In order to check the features of the proposed system, we test it with different interaction modalities (sudden impact, continuous pushing, multiple PoC) and in different situations (hovering, trajectory following).

Therefore, our aim is to show how different behaviors can be obtained using different tuning of the admittance controller, focusing on the stiffness constant S in equation (3.21). The other parameters are fixed for all experiments. In particular, the trajectory tracking controller parameters used for the translational dynamics in (3.22) are $\boldsymbol{K}_d = 5\boldsymbol{I}_3$, $\boldsymbol{K}_p = 25\boldsymbol{I}_3$ and $\boldsymbol{K}_i = 2\boldsymbol{I}_3$. Similarly $\boldsymbol{K}_\omega = 0.5\boldsymbol{I}_3$, $\boldsymbol{K}_r = 1.2\boldsymbol{I}_3$ and $\boldsymbol{K}_{ir} = 0.2\boldsymbol{I}_3$ were used

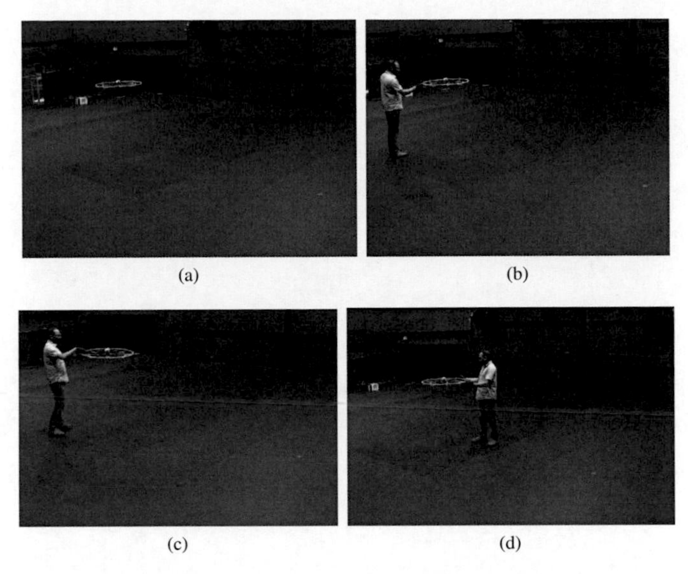

Figure 3.9: Snapshots of the human-UAV physical interaction experiment. 3.9(a): UAV in hovering state. 3.9(b): Human performing sudden impact during interaction. 3.9(c): Human performing continuous pushing during interaction. 3.9(d): Human performing multiple point of contact interaction.

for the rotational dynamics in (3.27). The Table. 3.1 lists all the experimental parameters used for human-UAV physical interaction.

3.8.1 Continuously Pushing, Sudden Impact and Multiple PoCs

In the first experiment, the quadrotor is allowed to hover at its take off position while different types of human interaction (continuous pushing, sudden impact and simultaneous multiple contacts) are performed as shown in Fig. 3.9. The plot of the sensor activation is shown in Fig. 3.8(a). Sudden impact contacts are applied for $t \in [25\,s, 42\,s]$, $t \in [58\,s, 68\,s]$, $t \in [74\,s, 94\,s]$ and $t \geq 105\,s$. In UAV-HRPI, sudden impacts may happen either voluntarily (i.e., an operator intentionally pushes away the UAV) or accidentally. In both cases, the UAV should be compliant in order to follow the commands of the operator or to go in the opposite direction with respect to where the contact happens. At $t = 44\,s$, $52\,s$ and $69\,s$, continuous interactions are performed at PoC positions s_2, s_6 and s_4 respectively. Multiple contacts are performed at $t = 99\,s$ with continuous interaction forces applied

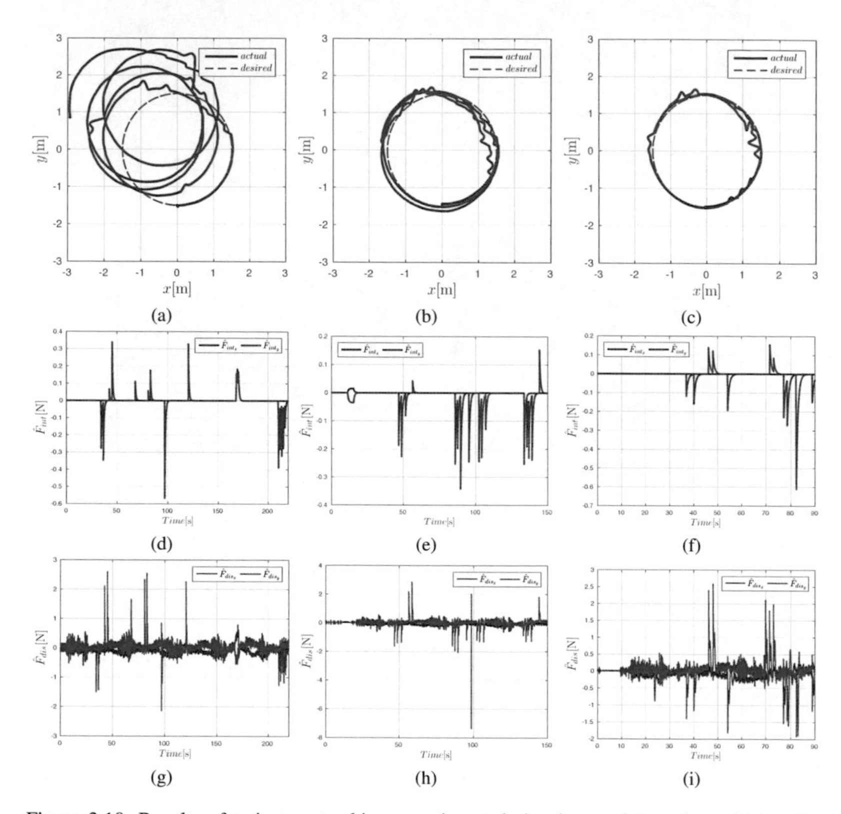

Figure 3.10: Results of trajectory tracking experiment during human interaction with varying stiffness parameter. 3.10(a-c): Two dimensional XY-plot of UAV tracking a circular trajectory during human interaction with stiffness constant 0.0, 0.1 and 0.5 respectively. 3.10(d-f):Estimated interaction force along X-axis ($\hat{\boldsymbol{F}}_{int_x}$) and Y-axis ($\hat{\boldsymbol{F}}_{int_y}$) with stiffness constant 0.0, 0.1 and 0.5 respectively. 3.10(g-i):Estimated disturbance along X-axis ($\hat{\boldsymbol{F}}_{dis_x}$) and Y-axis ($\hat{\boldsymbol{F}}_{dis_y}$) with stiffness constant 0.0, 0.1 and 0.5 respectively.

both at s_4 and s_6. These cases are compatible for example with a kinesthetic trajectory learning task, and also in this case the UAV should be completely compliant with the commands of the operator. Therefore, in this experiment we have tuned the admittance controller with stiffness $S = 0$.

Figures 3.8(c) and 3.8(d) show the estimated interaction forces $\hat{\boldsymbol{F}}_{int_x}$ and $\hat{\boldsymbol{F}}_{int_y}$ respectively. At the end of every human contact, the estimated interaction forces decrease slowly. This desirable behavior is due to the smoothening action of the 2nd-order low-pass filter which reduces the amplitude of the desired acceleration (Fig. 3.8(g) and Fig. 3.8(h)). Since the translational and rotational dynamics of the quadrotor are coupled (see equations (2.5)-(2.7)), reduced desired accelerations means also reduced desired pitch θ_d and roll ϕ_d, and smaller angular rates. Hence, the introduction of the low-pass filter helps the system in keeping a smooth, human-friendly behavior where no sudden acceleration or rotation is performed.

Figures 3.8(e) and 3.8(f) show the estimated external disturbance $\hat{\boldsymbol{F}}_{dis_x}$ and $\hat{\boldsymbol{F}}_{dis_y}$ respectively. At the end of every sudden impact, a spike in $\hat{\boldsymbol{F}}_{dis}$ is clearly visible. This happens because the dynamics of the UAV is heavily perturbed by a sudden impact, hence strong disturbance terms are estimated during and after the contact. However, a substantial fraction of the external force exerted by the user is still detected as interaction force, enough to detect the impact. As expected, during the continuous pushing scenarios this effect is either not present or negligible.

The reference trajectory produced by the admittance controller with the modified acceleration $\ddot{\boldsymbol{p}}_a$ (in red), velocity $\dot{\boldsymbol{p}}_a$ (in green) and position $\boldsymbol{p}_a$ (in blue) is shown in Fig. 3.8(g) and Fig. 3.8(h). The trajectory shows a smooth behavior thanks to the smoothing of the admittance force $\boldsymbol{F}_a$ performed by the low-pass filtering applied on $\hat{\boldsymbol{F}}_{int}$. Figure 3.8(b) shows the XY-plot of the position of the UAV during the experiment.

3.8.2 Trajectory tracking during human interaction with varying stiffness

In this experiment, the quadrotor is given a trajectory tracking task to follow a circle of $3\,m$ diameter (shown in blue in Fig. 3.10(a-c)) with a constant linear velocity of $0.3\,m/s$. Human contact is made at different PoC positions and $\boldsymbol{F}_{int}$ is applied to change the desired trajectory. Three experiments were performed with different stiffness constant values S while following the circular trajectory. In general, the value of the stiffness S decides the behavior of the quadrotor not only during the interaction phase, but also after the end of the interaction.

Initially, the experiment is carried out with stiffness constant $S = 0$. The XY-plot in Fig. 3.10(a) compares the actual (red) and reference (blue) trajectories. The estimated interaction forces $\hat{\boldsymbol{F}}_{int_x}$ and $\hat{\boldsymbol{F}}_{int_y}$ are shown in Fig. 3.10(d). The plots show that whenever a force is applied, its value is estimated and the trajectory is changed according to its magnitude. Hence, the original circular reference trajectory is modified to the new translated circular admittance trajectory $\boldsymbol{p}_a$ which is tracked by the UAV. The estimated $\hat{\boldsymbol{F}}_{dis_x}$ and $\hat{\boldsymbol{F}}_{dis_y}$ are shown in Fig. 3.10(g). This case is once again compatible with applications in which the UAV must follow the lead of the human. For example, if the UAV has to perform a prefixed trajectory (e.g., for monitoring, inspection, target tracking, data collection, etc.) in a place specified by an untrained operator.

The results of the same trajectory tracking task performed with stiffness $S = 0.1$ are shown in Figs. 3.10(b), 3.10(e) and 3.10(h). Also in this case the trajectory is modified by the presence of $\hat{\boldsymbol{F}}_{int_x}$ and $\hat{\boldsymbol{F}}_{int_y}$. However, thanks to the nonzero stiffness in the admittance control, every time the UAV is able to come back to follow the original desired circular trajectory. The convergence speed is nevertheless very slow, and it takes more than one circle for the robot to recover the initial trajectory.

In the third test (Figs. 3.10(c), 3.10(f) and 3.10(i)), the same circular trajectory tracking task is carried out with $S = 0.5$. Due to the larger stiffness in the admittance control, the controller acts immediately on the position error generating a $\boldsymbol{p}_a$ which tries to bring the quadrotor back to the original trajectory as fast as possible. Hence, each interacting force leaves only a small bump in the trajectory (Fig. 3.10(c)).

In both cases in which the stiffness is greater than zero, the UAV shows a compliant behavior but also goes back to the original trajectory. Hence, they are compatible with a situation in which the UAV is in charge of delivering one or more objects (e.g., tools) to one or more humans without landing. In fact, in such situation the UAV should show some compliance in order to facilitate the pick up operation, but should also come back to the prefixed trajectory in order to continue the task after the pick up from one of the recipient. The stiffness could be set higher or lower depending on the type and length of the trajectory, the distance between the operators and the type of object that should be picked up. Additional studies can be carried out in order to determine the optimal stiffness to facilitate the pick up operation.

3.9 Discussions and Possible Extensions

Very few researchers have started looking in the direction of *Human-UAV physical interaction*, though *UAV manipulation and interaction with the environment* is one of the important research domain in aerial robotics. In this chapter major hardware improvements and interaction methodologies have been proposed. Summarizing our contribution:

1. it is presented a novel hardware setup for UAV-HRPI (Sec. 3.4) that offers the possibility to exchange, detect and characterize interaction forces between humans and UAVs. The sensor ring provides a safe interaction surface for humans and push button data help to characterize the external wrench by separating interaction forces from the disturbances;

2. after modeling and estimating the external wrench acting on a UAV as the sum of interaction forces and disturbances, it is developed a methodology (Sec. 3.5)

to separate them based on residual estimation and quadratic programming. This methodology utilizes the data from the push button sensor and is computationally fast confirming that it can be used in real-time UAV application;

3. the estimated interaction forces are used in an admittance control paradigm (see Sec. 3.6.1) which modifies the desired trajectory on the basis of the human interaction by modifying the physical properties of the UAV considering them as mass-spring-damper system. The estimated disturbances are instead rejected through a modified geometric tracking controller (Sec. 3.6.2) to track the reference trajectory provided by the admittance controller;

4. extensive simulations (Sec. 3.7) and experiments (Sec. 3.8) have been conducted assuming different scenarios as continuous human contact, sudden impact and multiple points of contact. The results validates the effectiveness of our approach. Experiments also show the UAV behavior and possibility to vary the stiffness parameter depending on the application scenario.

Some improvements can be done to our current setup. Although providing an interaction surface already account for some form of safety, more safety feature could be added by improving the hardware design to offer a larger 3D surface which fully encloses the propellers with full protection for the user. This would ensure that the UAV is foolproof safe during human interaction. A shortcoming of the proposed estimation approach is when both interaction and disturbance wrenches occur in the same direction. Further studies could be done to address this limitation, e.g., by developing an adaptive law for the weight in eq. ((3.12)). Additionally, sensors can also be deployed three-dimensionally which would allow the identification of the point of contact in all directions.

Here, the disturbance rejection and the trajectory tracking is handled by a geometric tracking controller suitably modified to include a feedforward disturbance term. This utilizes the best state estimates available from the indoor motion capture system. However, employing a robust non-linear controller which can work effectively against the unknown system parameters, uncertainties and external perturbations has been proved to be useful for outdoor applications, when the environment is unknown with lot of uncertainties. This idea will be expanded in Chapter. 4.

Though the exchange of forces between humans and UAVs is possible in this UAV-HRPI setup, the exchange of intuitive torques with the UAV is infeasible because of the under-actuation of the quadrotor UAV in its system dynamics. The torque exchange between humans and UAVs could be addressed only when this problem is solved through a fully actuated UAV, which would then allow the human to command also the rotational degrees of freedom. The safety concern of UAV-HRPI can also be better addressed with a fully actuated system, whose development is addressed in Chapter. 5.

Apart from the engineering technological advancements, in the scientific direction, this platform can also be used for studies with respect to human subjects to study interaction behavior patterns with UAVs.

Chapter 4

Robust Adaptive Super Twisting Control

Is there a need for a robust controller when the literature already have many controllers proved to be effective for UAV applications? Is there a control design which can effectively act against the parameter uncertainties, unmodeled dynamics and unknown disturbances? Is there a controller which would be effective in an unstructured outdoor environment?

In this chapter we try to answer all these questions by means of proposing a non-linear adaptive super twisting controller for quadrotor UAV. The quadrotor system dynamics are raised to higher differential order so that there is a input-output decoupling available and the control design could be effectively applied.

The discussion presented in this chapter is based upon the work published in (Rajappa *et al.*, 2016).

4.1 Introduction

In the past decade we have witnessed the blooming of aerial robotics as a research domain and commercial success. Unmanned Aerial Vehicles (UAVs) are increasingly used in industrial and civilian applications because their mobility makes them capable to access dangerous areas both in indoor and outdoor scenarios and to tackle a wide variety of tasks. Initially the focus of research was on navigation tasks (mapping, surveillance, etc.), but recently the attention has shifted more towards physical interaction with the environment and manipulation of objects (Orsag *et al.*, 2013),(Lippiello and Ruggiero, 2012), (Pounds *et al.*, 2011; Lindsey *et al.*, 2011), (Gioioso *et al.*, 2014b). There are many linear controllers (PD control) available which are reliable for UAV applications when the state estimation, system model and the environment are well known or defined. However as the uncertainty space starts to grow with the different applications with UAV, the effectiveness of these controllers are always questioned. During a human-UAV physical interaction scenario (as discussed in Chapter. 3), it absolutely becomes a necessity to have a robust control.

From this shift towards physical interaction it is emerging the need to have not only good accuracy in trajectory tracking but also robustness to perturbations, such as external disturbances (e.g.: wind gusts) and model uncertainties (e.g.: change in mass when grasping an object).

4.1.1 Related Works

Indeed, classical control approaches such as nonlinear dynamic inversion and feedback linearization (Mistler *et al.*, 2001) are known for their vulnerability to model uncertainties (Lee *et al.*, 2009). Sliding mode control (Derafa *et al.*, 2010) appears to be a promising solution to deal with model uncertainties because it has well known perturbations rejection properties (Shtessel *et al.*, 2014). Yet, this control strategy is also known to suffer from chattering which might reduce the performance and degrade the actuators. In order to overcome this problem, adaptive sliding mode strategies have been proposed both for quadrotors (Bouadi *et al.*, 2011) and for fixed wing aircrafts (Castaneda *et al.*, 2013).

All the cited control approaches have some limitations. While the controller proposed in (Bouadi *et al.*, 2011) still suffers from chattering, (Lee *et al.*, 2009) shows limited robustness to parameter uncertainties, and (Derafa *et al.*, 2010) requires the knowledge of the upper bound on perturbations which in most practical cases is impossible to estimate, therefore leading to over-conservative tunings. The most commonly used near-hovering controller (Lee *et al.*, 2013) and geometric tracking controller (Lee *et al.*, 2010) are vulnerable in outdoor conditions when there are extreme perturbations. Moreover situations of aggressive maneuvers expects more robust performance. Another non-linear controller proposed by Liu *et al.* (2015) for nontrivial maneuvers requires the knowledge of all parameters for robust performance.

4.1.2 Methodologies

Taking into account all the requirements and the limitations in the earlier control methodologies mentioned in Sec. 4.1.1, in this chapter,

1. it is implemented an Adaptive Super Twisting Controller (ASTC) with the following properties:

    - It considers and compensates for all the uncertainties (parametric, model, disturbances) lumped together.
    - It does not require any knowledge of the upper bound of the uncertainties.
    - It adapts the gains rather than the model parameter (Shtessel *et al.*, 2012). In this way, the gains are lowered whenever possible, thus reducing control actions, chattering and noise amplification.

- It uses a feedforward dynamic inversion (FF) to reduce the discontinuous control, thus improving performance and further reducing chattering.

2. it is derived the regular control form of the quadrotor system in the state space model. The regular control form includes taking the translation dynamics to a higher order through double differentiation;

3. it is modeled the system parametric uncertainties and disturbances as a part of the lumped perturbations;

4. the controller is validated by means of physical simulation as well as compared with the standard super twisting controller which doesn't have the gain adaptation facility to see the improvement in the performance.

4.2 Preliminary System Descriptions

The quadrotor UAV model for which the robust controller is designed and validated is the one that has been introduced earlier in Sec. 1.2 and whose system dynamics are defined in Sec. 2.2 of Chapter. 2. Here in this section after a brief introduction of the quadrotor dynamics in state-space model, it is redefined in regular control form which is then used for the controller design. Then it is modeled the parameter uncertainty and external perturbations.

4.2.1 Dynamic System Model

The reference inertial frame $\mathcal{F}_W$, body frame $\mathcal{F}_B$ and the notations for the quadrotor model follows the same philosophy as in Sec. 2.2 seen in Fig. 2.1. The basic quadrotor states $\boldsymbol{\xi}_W$ defined in $\mathcal{F}_W$ is as in (2.1) and the generalized velocity vector states $\boldsymbol{\zeta}$ are expressed as defined in (2.4). The matrix $\boldsymbol{R}_B^W$, which represent the rotation between $\mathcal{F}_B$ and $\mathcal{F}_W$ is defined in (2.2). The dynamical model of the quadrotor representing the translational and rotational dynamics expressed using the Newton-Euler formulation in $\mathcal{F}_W$ is given by (2.5)-(2.7). For the sake of continuity in this chapter, the dynamic model are mentioned again as,

$$m\ddot{\boldsymbol{p}}_W = -mge_3 + \rho \boldsymbol{R}_B^W e_3 + \boldsymbol{R}_H^W \boldsymbol{F}_{ext} \,, \tag{4.1}$$

$$\boldsymbol{I}_B \dot{\boldsymbol{\omega}}_B = -\boldsymbol{\omega}_B \times \boldsymbol{I}_B \boldsymbol{\omega}_B + \boldsymbol{\tau} + \boldsymbol{\tau}_{ext} \,, \tag{4.2}$$

$$\dot{\boldsymbol{\Theta}}_W = \boldsymbol{T}(\boldsymbol{\Theta}_W)\boldsymbol{\omega}_B \,. \tag{4.3}$$

The standard transformation matrix $\boldsymbol{T}(\boldsymbol{\Theta}_W)$ from $\boldsymbol{\omega}_B$ to the Euler angle rates $\dot{\boldsymbol{\Theta}}_W$ is given by (2.9). Refer Sec. 2.2 for details of the individual variables in (4.1)–(4.3). Here the translational dynamics of the quadrotor are expressed in $\mathcal{F}_W$, while the rotational dynamics are expressed in $\mathcal{F}_B$ which have been proved to be convenient for controller design.

The dynamic model mentioned in (4.1)–(4.3), can be written in state-space form as:

$$\dot{\boldsymbol{x}} = \boldsymbol{f}(\boldsymbol{x}) + \boldsymbol{g}(\boldsymbol{x})\boldsymbol{u} \tag{4.4}$$

where the system states

$$\boldsymbol{x} = \begin{bmatrix} x & y & z & \phi & \theta & \psi & \dot{x} & \dot{y} & \dot{z} & p & q & r \end{bmatrix}^T \in \mathbb{R}^{12\times 1}, \tag{4.5}$$

$$\boldsymbol{f}(\boldsymbol{x}) = \begin{bmatrix} \dot{x} \\ \dot{y} \\ \dot{z} \\ f_{(4,1)} \\ f_{(5,1)} \\ f_{(6,1)} \\ 0 \\ 0 \\ g \\ \frac{Iyy-Izz}{Ixx}qr \\ \frac{Izz-Ixx}{Iyy}pr \\ \frac{Ixx-Iyy}{Izz}pq \end{bmatrix}, \boldsymbol{g}(\boldsymbol{x}) = \begin{bmatrix} 0 & 0 & 0 & 0 \\ 0 & 0 & 0 & 0 \\ 0 & 0 & 0 & 0 \\ 0 & 0 & 0 & 0 \\ 0 & 0 & 0 & 0 \\ 0 & 0 & 0 & 0 \\ g_{(7,1)} & 0 & 0 & 0 \\ g_{(8,1)} & 0 & 0 & 0 \\ g_{(9,1)} & 0 & 0 & 0 \\ 0 & \frac{1}{Ixx} & 0 & 0 \\ 0 & 0 & \frac{1}{Iyy} & 0 \\ 0 & 0 & 0 & \frac{1}{Izz} \end{bmatrix}, \tag{4.6}$$

$$\boldsymbol{u} = \begin{bmatrix} u_1 \\ u_2 \\ u_3 \\ u_4 \end{bmatrix} = \begin{bmatrix} \rho \\ \tau_x \\ \tau_y \\ \tau_z \end{bmatrix} = \begin{bmatrix} b(\Omega_1^2 + \Omega_2^2 + \Omega_3^2 + \Omega_4^2) \\ bl(\Omega_4^2 - \Omega_2^2) \\ bl(\Omega_3^2 - \Omega_1^2) \\ d(\Omega_2^2 + \Omega_4^2 - \Omega_1^2 - \Omega_3^2) \end{bmatrix}, \tag{4.7}$$

with

$$\begin{cases} f_{(4,1)} = p + q\sin\phi\tan\theta + r\cos\phi\tan\theta \\ f_{(5,1)} = q\cos\phi - r\cos\phi \\ f_{(6,1)} = q\sin\phi\sec\theta + r\cos\phi\sec\theta \\ g_{(7,1)} = -\frac{1}{m}(\cos\phi\cos\psi\sin\theta + \sin\phi\sin\psi) \\ g_{(8,1)} = -\frac{1}{m}(\cos\phi\sin\psi\sin\theta - \sin\phi\cos\psi) \\ g_{(9,1)} = -\frac{1}{m}(\cos\phi\cos\theta). \end{cases}$$

Here, the control input in (4.7) is obtained from the relationship between the propeller velocities $\Omega_i, i \in 1 \rightarrow 4$ and the generated wrench (force/torque) given in (2.16). In order to ensure that the resulting Ω_i is feasible, we make the following assumption:

Assumption 4.1:
The control input is bounded, i.e., $\boldsymbol{u} \in \mathcal{U} = \{\boldsymbol{u}^{\star} \in [\boldsymbol{u}_{min}, \boldsymbol{u}_{max}]\}$.

Since the control input in (4.7) is related to the speed of the propellers, Assumption 4.1 implies that the speed of the propellers is always feasible through limiting the propeller angular velocities Ω_i.

4.2.2 Regular Control Form

As already mentioned in Sec. 1.2, the quadrotor is an underactuated system with only four control inputs. Therefore the main objective here is to control the absolute position of the UAV, $\boldsymbol{p}_d(t) = \begin{bmatrix} x_d & y_d & z_d \end{bmatrix}^T$ and the yaw angle ψ_d. Hence the output function of the control problem is chosen as

$$\boldsymbol{y} = \boldsymbol{h}(\boldsymbol{x}) = \begin{bmatrix} x & y & z & \psi \end{bmatrix}^T. \tag{4.8}$$

Let the relative degree $r_i, i \in 1 \to 4$, be defined as the number of times the ith output is differentiated until we arrive at the control input explicitly appearing. Utilizing (4.1)–(4.3), we obtain

$$\begin{aligned} \begin{bmatrix} \ddot{x} \\ \ddot{y} \\ \ddot{z} \\ \ddot{\psi} \end{bmatrix} &= a(\boldsymbol{x}) + b(\boldsymbol{x})\,\boldsymbol{u} \\ &= \begin{bmatrix} 0 \\ 0 \\ g \\ \frac{Ixx-Iyy}{Izz} pq \end{bmatrix} + \begin{bmatrix} g_{(7,1)} & 0 & 0 & 0 \\ g_{(8,1)} & 0 & 0 & 0 \\ g_{(9,1)} & 0 & 0 & 0 \\ 0 & 0 & 0 & \frac{1}{Izz} \end{bmatrix} \boldsymbol{u} = \boldsymbol{v}\,. \end{aligned} \tag{4.9}$$

Here $\boldsymbol{v} = \begin{bmatrix} v_1 & v_2 & v_3 & v_4 \end{bmatrix}^T$ is the external virtual control input introduced to solve the control problem. The non-linear system defined in (4.4) is not solvable by

$$\boldsymbol{u} = \alpha(\boldsymbol{x}) + \beta(\boldsymbol{x})\,\boldsymbol{v} \tag{4.10}$$

where $\alpha(\boldsymbol{x}) = -b(\boldsymbol{x})^{-1} a(\boldsymbol{x})$ and $\beta(\boldsymbol{x}) = b(\boldsymbol{x})^{-1}$, because $b(\boldsymbol{x})$ as seen from (4.9) is singular and not invertible. This can be explained from the derivatives $\ddot{x}$, $\ddot{y}$ and $\ddot{z}$, which are all affected by u_1 but none of them is affected by u_2, u_3, u_4. In order to get $b(\boldsymbol{x})$ non-singular, it would make sense to render $\ddot{x}$, $\ddot{y}$ and $\ddot{z}$ independent of u_1, by taking higher order derivatives and relating with the appearing other control inputs u_2, u_3, u_4. Therefore, it is clear that the non-interacting control solution can be provided by input-output system model decoupling through dynamic feedback linearizing with the output defined in (4.8).

First, let us introduce a new control input, i.e, $\bar{\boldsymbol{u}}$

$$\bar{\boldsymbol{u}} = \begin{bmatrix} \ddot{u}_1 \\ u_2 \\ u_3 \\ u_4 \end{bmatrix} = \begin{bmatrix} \bar{u}_1 \\ u_2 \\ u_3 \\ u_4 \end{bmatrix} \tag{4.11}$$

obtained by considering a dynamic extension of (4.4). Here $\bar{u}_1$ is obtained by the double differentiation of u_1 as,

$$u_1 = \rho, \tag{4.12a}$$
$$\dot{\rho} = \varsigma, \tag{4.12b}$$
$$\dot{\varsigma} = \bar{u}_1. \tag{4.12c}$$

The new extended system will have the form

$$\dot{\bar{\boldsymbol{x}}} = \boldsymbol{f}(\bar{\boldsymbol{x}}) + \boldsymbol{g}(\bar{\boldsymbol{x}})\,\bar{\boldsymbol{u}} \tag{4.13}$$

where the extended state is

$$\bar{\boldsymbol{x}} = \begin{bmatrix} x & y & z & \phi & \theta & \psi & \dot{x} & \dot{y} & \dot{z} & \rho & \varsigma & p & q & r \end{bmatrix}^T \in \mathbb{R}^{14\times 1} \tag{4.14}$$

and

$$\boldsymbol{f}(\bar{\boldsymbol{x}}) = \begin{bmatrix} \dot{x} \\ \dot{y} \\ \dot{z} \\ f_{(4,1)} \\ f_{(5,1)} \\ f_{(6,1)} \\ g_{(7,1)}\rho \\ g_{(8,1)}\rho \\ g + g_{(9,1)}\rho \\ \varsigma \\ 0 \\ \frac{Iyy-Izz}{Ixx}qr \\ \frac{Izz-Ixx}{Iyy}pr \\ \frac{Ixx-Iyy}{Izz}pq \end{bmatrix}, \boldsymbol{g}(\bar{\boldsymbol{x}}) = \begin{bmatrix} 0 & 0 & 0 & 0 \\ 0 & 0 & 0 & 0 \\ 0 & 0 & 0 & 0 \\ 0 & 0 & 0 & 0 \\ 0 & 0 & 0 & 0 \\ 0 & 0 & 0 & 0 \\ 0 & 0 & 0 & 0 \\ 0 & 0 & 0 & 0 \\ 0 & 0 & 0 & 0 \\ 0 & 0 & 0 & 0 \\ 1 & 0 & 0 & 0 \\ 0 & \frac{1}{Ixx} & 0 & 0 \\ 0 & 0 & \frac{1}{Iyy} & 0 \\ 0 & 0 & 0 & \frac{1}{Izz} \end{bmatrix}. \tag{4.15}$$

Considering the output vector $\mathbf{y} = \begin{bmatrix} x & y & z & \psi \end{bmatrix}^T$ as defined in (4.8), it is easy to see that the relative degree for all the four components of the output is $r_1 = r_2 = r_3 = 4$ and $r_4 = 2$, i.e., $\sum_{i=1}^{4} r_i = 14$, which is equal to the number of the extended system states as seen in (4.14). Therefore, there exists a diffeomorphism $\Phi(\bar{\boldsymbol{x}})$ through which the system can be transformed via dynamic feedback using the suitable change of coordinates into a fully linear and controllable one (Isidori, 1995). Therefore the coordinates transformation $\boldsymbol{z} = \Phi(\bar{\boldsymbol{x}})$ defined by

$$\begin{cases} z_1 = x, \quad z_2 = \dot{x}, \quad z_3 = \ddot{x}, \quad z_4 = \dddot{x}, \\ z_5 = y, \quad z_6 = \dot{y}, \quad z_7 = \ddot{y}, \quad z_8 = \dddot{y}, \\ z_9 = z, \quad z_{10} = \dot{z}, \quad z_{11} = \ddot{z}, \quad z_{12} = \dddot{z}, \\ \qquad z_{13} = \psi, \quad z_{14} = \dot{\psi} \end{cases} \tag{4.16}$$

transforms (4.13) into a regular form in which the dynamics of the output $\boldsymbol{y}$ in (4.8) are decoupled into a chain of integrators. The system transformation with the new states $z = [z_1, z_2, \ldots.., z_{14}]^T$ can be written in state-space form as

$$\dot{\boldsymbol{z}} = \begin{bmatrix} z_2 \\ z_3 \\ z_4 \\ a_x(z) \\ z_6 \\ z_7 \\ z_8 \\ a_y(z) \\ z_{10} \\ z_{11} \\ z_{12} \\ a_z(z) \\ z_{14} \\ a_\psi(z) \end{bmatrix} + \begin{bmatrix} \mathbf{0}_{3\times4} \\ b_x(z) \\ \mathbf{0}_{3\times4} \\ b_y(z) \\ \mathbf{0}_{3\times4} \\ b_z(z) \\ \mathbf{0}_{1\times4} \\ b_\psi(z) \end{bmatrix} \begin{bmatrix} \bar{u}_1 \\ u_2 \\ u_3 \\ u_4 \end{bmatrix}, \tag{4.17}$$

where

$$\begin{cases} \begin{bmatrix} a_x(z) \\ a_y(z) \\ a_z(z) \end{bmatrix} = \underbrace{\begin{pmatrix} -\frac{\rho S(\boldsymbol{R}_\Theta e_3)\boldsymbol{R}_\Theta \boldsymbol{I}_B^{-1} S(\omega)\boldsymbol{I}_B\omega}{m} \\ -\frac{\boldsymbol{R}_\Theta S(\dot{\boldsymbol{\Theta}})^2 e_3 \rho}{m} - \frac{2\boldsymbol{R}_\Theta S(\dot{\boldsymbol{\Theta}}) e_3 \varsigma}{m} \end{pmatrix}}_{3\times1} \\ \begin{bmatrix} b_x(z) \\ b_y(z) \\ b_z(z) \end{bmatrix} = \begin{bmatrix} \underbrace{-\frac{\boldsymbol{R}_\Theta e_3}{m}}_{3\times1} & \underbrace{\frac{\rho S(\boldsymbol{R}_\Theta e_3)\boldsymbol{R}_\Theta \boldsymbol{I}_B^{-1}}{m}}_{3\times3} \end{bmatrix} \\ a_\psi(z) \quad = [\dot{\boldsymbol{T}}(\Theta)\omega - \boldsymbol{T}(\Theta)\boldsymbol{I}_B^{-1} S(\omega)\boldsymbol{I}_B\omega]_3 \\ b_\psi(z) \quad = \begin{bmatrix} \mathbf{0}_{3\times1} & \underbrace{\boldsymbol{T}(\Theta)\boldsymbol{I}_B^{-1}}_{3\times3} \end{bmatrix}_3 . \end{cases} \tag{4.18}$$

Here $S(\omega)$ is the skew-symmetric matrix of ω such that $\dot{\boldsymbol{R}_{\Theta}} = \boldsymbol{R}_{\Theta} S(\omega)$ and $S(\dot{\boldsymbol{\Theta}})$ is the skew-symmetric matrix of $\dot{\boldsymbol{\Theta}}$ which describe the Euler angle rates in $\mathcal{F}_W$. The subscript 3 in $a_{\psi}(z)$ and $b_{\psi}(z)$ means that only the third row of the expression is selected. The system equations expressed in z and $\bar{\boldsymbol{u}}$ are

$$\begin{bmatrix} \dddot{p} \\ \ddot{\psi} \end{bmatrix} = \begin{bmatrix} \dddot{x} \\ \dddot{y} \\ \dddot{z} \\ \ddot{\psi} \end{bmatrix} = \begin{bmatrix} \dot{z}_4 \\ \dot{z}_8 \\ \dot{z}_{12} \\ \dot{z}_{14} \end{bmatrix} = \underbrace{\begin{bmatrix} a_x(z) \\ a_y(z) \\ a_z(z) \\ a_{\psi}(z) \end{bmatrix}}_{\triangleq \boldsymbol{a}(z)} + \underbrace{\begin{bmatrix} b_x(z) \\ b_y(z) \\ b_z(z) \\ b_{\psi}(z) \end{bmatrix}}_{\triangleq \boldsymbol{b}(z)} \bar{\boldsymbol{u}}. \tag{4.19}$$

As clear from (4.16), the state of the new system includes the jerk, which in general is not directly measurable. Therefore, for control purpose it can be computed as

$$\dddot{p} = -\frac{1}{m} \left(\boldsymbol{R}_{\Theta} S(\omega) e_3 \rho + \boldsymbol{R}_{\Theta} e_3 \varsigma \right), \tag{4.20}$$

which is obtained through differentiation of the translational dynamics in (4.1). For the new system model representation given by (4.19) to hold, we take the following assumption:

Assumption 4.2:
The roll and pitch angles ϕ and θ are limited to $(-\pi/2,\ \pi/2)$.

Assumption 4.2 ensures that the matrix $\boldsymbol{b}(z)$ in (4.19) is non-singular, because $\boldsymbol{T}(\Theta)$ in (4.18) is non-singular, and always has $rank(\boldsymbol{b}(z)) = 4$, therefore being invertible.

4.2.3 Uncertainties

The model presented in (4.19) depicts the system without uncertainties. To incorporate the effect of inexact knowledge of the parameters and of disturbances, we consider that:

1. the quadrotor is subject to external disturbances $\boldsymbol{\chi}$ that act w.r.t. the CoM as force and torque wrenches. The dynamic equation (4.19) becomes

$$\begin{bmatrix} \dddot{p} \\ \ddot{\psi} \end{bmatrix} = \boldsymbol{a}(z) + \boldsymbol{b}(z)(\bar{\boldsymbol{u}} + \boldsymbol{\chi}); \tag{4.21}$$

2. only the dynamic parameters m, $\boldsymbol{I}_B$ are uncertain.

Following these assumptions the above model (4.21) becomes

$$\begin{bmatrix} \dddot{p} \\ \ddot{\psi} \end{bmatrix} = \boldsymbol{a}_n + \Delta\boldsymbol{a} + \boldsymbol{b}_n(\bar{\boldsymbol{u}} + \boldsymbol{\chi}) + \Delta\boldsymbol{b}(\bar{\boldsymbol{u}} + \boldsymbol{\chi}) = \\ = \boldsymbol{a}_n + \boldsymbol{b}_n\boldsymbol{u} + \boldsymbol{\kappa}\,, \tag{4.22}$$

where

- $\boldsymbol{a}_n$ and $\boldsymbol{b}_n$ describe the nominal model of the robot;
- $\Delta\boldsymbol{a}$ and $\Delta\boldsymbol{b}$ contain the parametric uncertainties;
- $\boldsymbol{\kappa} = \boldsymbol{b}_n\boldsymbol{\chi} + \Delta\boldsymbol{a} + \Delta\boldsymbol{b}(\boldsymbol{u} + \boldsymbol{\chi})$ is the vector of lumped perturbations.

Note that $\boldsymbol{b}_n$ is always full rank (Assumption 4.2), so the lumped perturbations satisfy the matching condition. Moreover, we make an additional assumption:

Assumption 4.3:
$\boldsymbol{\kappa}$ is bounded as $\|\boldsymbol{\kappa}\|_2 \leq \kappa_{\max}$, but the bound $\kappa_{\max} \geq 0$ is unknown.

In practice it is difficult to estimate the upper bound on $\boldsymbol{\kappa}$. This could lead to over-conservative gain tuning and consequently to unnecessary high control actions, chattering and noise amplification. Finally, we want to underline that we consider the case that only the dynamic parameters are uncertain.

4.3 Control

In this section we propose our solution for trajectory tracking using a quadrotor in the presence of the lumped disturbance $\boldsymbol{\kappa}$. The trajectory is specified as a desired position $\boldsymbol{p}_d(t) = \begin{bmatrix} x_d & y_d & z_d \end{bmatrix}^T$ with its derivatives up to the snap $\ddddot{p}_d(t)$, and desired yaw ψ_d and its derivatives up to the second order $\ddot{\psi}_d$. Such a trajectory can be easily defined offline or computed online using input shaping or filtering techniques. We assume that the state variables defined in (4.16) are available at every time instant.

The tracking controller is designed as a robust law $\bar{\boldsymbol{u}}$ of the form

$$\bar{\boldsymbol{u}} = \bar{\boldsymbol{u}}_{sm} + \bar{\boldsymbol{u}}_{ff}, \tag{4.23}$$

where

- $\bar{\boldsymbol{u}}_{sm}$ is a term based on a sliding mode approach;
- $\bar{\boldsymbol{u}}_{ff}$ is a feedforward term based on the dynamic inversion of the nominal model.

In order to compute $\boldsymbol{u}$ from $\bar{\boldsymbol{u}}$, we need to double integrate u_1. In the remaining of this section we detail the two terms that compose $\bar{\boldsymbol{u}}$ in (4.23).

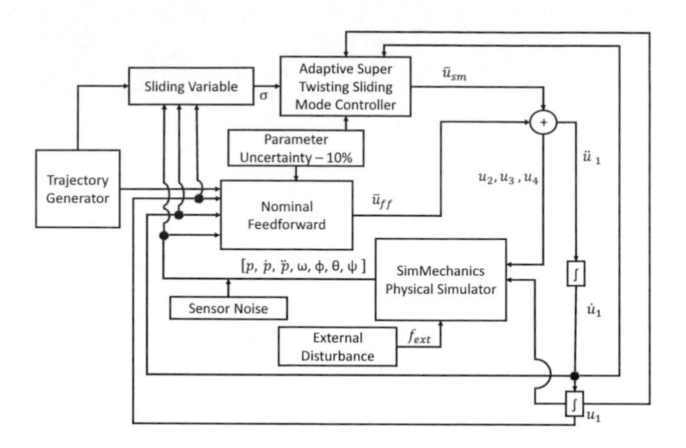

Figure 4.1: Control scheme architecture. Jerk ($\dddot{\boldsymbol{p}}$) and snap ($\ddddot{\boldsymbol{p}}$) required in the adaptive controller and nominal feedback are calculated from acceleration ($\ddot{\boldsymbol{p}}$) of the quadrotor.

4.3.1 Adaptive Super Twisting Control

The sliding mode control term $\bar{\boldsymbol{u}}_{sm}$ is designed to steer to zero the tracking errors of position $\boldsymbol{e_p} = \boldsymbol{p} - \boldsymbol{p}_d = \begin{bmatrix} e_x & e_y & e_z \end{bmatrix}^T \in \mathbb{R}^3$ and yaw error $e_\psi = \psi - \psi_d$ in presence of the uncertainties $\boldsymbol{\kappa}$. As seen earlier in Sec. 4.2.2, in the model in regular form (4.19), the output is decoupled. Therefore, the sliding variable is chosen as

$$\boldsymbol{\sigma} = \begin{bmatrix} \sigma_x \\ \sigma_y \\ \sigma_z \\ \sigma_\psi \end{bmatrix} = \begin{bmatrix} \dddot{e}_x + \lambda_{x_3}\ddot{e}_x + \lambda_{x_2}\dot{e}_x + \lambda_{x_1}e_x \\ \dddot{e}_y + \lambda_{y_3}\ddot{e}_y + \lambda_{y_2}\dot{e}_y + \lambda_{y_1}e_y \\ \dddot{e}_z + \lambda_{z_3}\ddot{e}_z + \lambda_{z_2}\dot{e}_z + \lambda_{z_1}e_z \\ \dot{e}_\psi + \lambda_{\psi_1}e_\psi \end{bmatrix}, \tag{4.24}$$

where $\lambda \in \mathbb{R}^{n\times n}$ is a positive definite diagonal matrix. Using (4.19), the time derivative of $\boldsymbol{\sigma}$ is

$$\dot{\boldsymbol{\sigma}} = \begin{bmatrix} -\ddddot{x}_d + \lambda_{x_3}\dddot{e}_x + \lambda_{x_2}\ddot{e}_x + \lambda_{x_1}\dot{e}_x \\ -\ddddot{y}_d + \lambda_{y_3}\dddot{e}_y + \lambda_{y_2}\ddot{e}_y + \lambda_{y_1}\dot{e}_y \\ -\ddddot{z}_d + \lambda_{z_3}\dddot{e}_z + \lambda_{z_2}\ddot{e}_z + \lambda_{z_1}\dot{e}_z \\ -\ddot{\psi}_d + \lambda_{\psi_1}\dot{e}_\psi \end{bmatrix} + \boldsymbol{a}(\boldsymbol{z}) + \boldsymbol{b}(\boldsymbol{z})\,\bar{\boldsymbol{u}} \tag{4.25}$$

showing that $\boldsymbol{\sigma}$ has relative degree one with respect to $\bar{\boldsymbol{u}}$. To achieve the 2-sliding mode $\boldsymbol{\sigma} = \dot{\boldsymbol{\sigma}} = \boldsymbol{0}$, we implement $\bar{\boldsymbol{u}}_{sm}$ according to the well known Super Twisting controller

(STC) (Shtessel *et al.*, 2014; Levant, 1993). The expression of the standard STC is

$$\begin{aligned}\bar{\boldsymbol{u}}_{sm} &= \boldsymbol{b}(\boldsymbol{z})^{-1}\left(-\boldsymbol{\alpha}\,|\boldsymbol{\sigma}|^{\frac{1}{2}}\,sign(\boldsymbol{\sigma})+\boldsymbol{v}\right)\\ \dot{\boldsymbol{v}} &= \begin{cases}-\bar{\boldsymbol{u}}_{sm} & \text{if}\,|\bar{\boldsymbol{u}}_{sm}|>\bar{\boldsymbol{u}}_m\\ -\boldsymbol{\beta}\,sign(\boldsymbol{\sigma}) & \text{if}\,|\bar{\boldsymbol{u}}_{sm}|\le\bar{\boldsymbol{u}}_m\end{cases}.\end{aligned} \tag{4.26}$$

Here, $\bar{\boldsymbol{u}}_m$ denotes an upper bound for $\bar{\boldsymbol{u}}_{sm}$ and $\boldsymbol{\alpha}$, $\boldsymbol{\beta}$ are definite positive diagonal matrices of gains. The control law (4.26) has two remarkable properties, i) it does not require the knowledge of $\dot{\boldsymbol{\sigma}}$ and therefore of the snap $\ddot{\ddot{\boldsymbol{p}}}$ and yaw acceleration $\ddot{\psi}$, and ii) the discontinuous function $sign(\boldsymbol{\sigma})$ is integrated, thus significantly attenuating chattering.

From (Shtessel *et al.*, 2014) it is proved that the standard STC controller achieves finite-time convergence to the 2^{nd}order-sliding manifold with few assumptions. In particular, it is necessary to choose the gains $\boldsymbol{\alpha}$ and $\boldsymbol{\beta}$ high enough, according to the upper bound on $\boldsymbol{\kappa}$. Since the upper bound on $\boldsymbol{\kappa}$ is not known (Assumption 4.3) we adapt the gains online according to the law proposed in (Shtessel *et al.*, 2010, 2012),

$$\begin{aligned}\dot{\boldsymbol{\alpha}} &= \begin{cases}\omega_{\alpha}\sqrt{\frac{\gamma}{2}}sign(|\boldsymbol{\sigma}|-\boldsymbol{\mu}), & \text{if}\ \boldsymbol{\alpha}>\boldsymbol{\alpha}_m\\ \eta, & \text{if}\ \boldsymbol{\alpha}\le\boldsymbol{\alpha}_m\end{cases}\\ \boldsymbol{\beta} &= 2\varepsilon\,\boldsymbol{\alpha}\,,\end{aligned} \tag{4.27}$$

where

- $\omega_{\alpha}, \gamma, \eta$ are arbitrary positive constants;
- $\boldsymbol{\alpha}_m$ is an arbitrary small positive constant introduced to keep the gains positive;
- $\boldsymbol{\mu}$ is a positive parameter that defines the boundary layer for the real sliding mode.

Under few mild assumptions (Shtessel *et al.*, 2012), the STC with adaptive gains (4.27) achieves finite-time convergence to a real 2-sliding mode $\|\boldsymbol{\sigma}\|\le\boldsymbol{\mu}_1$ and $\|\boldsymbol{\sigma}\|\le\boldsymbol{\mu}_2$, with $\boldsymbol{\mu}_1\ge\boldsymbol{\mu}$ and $\boldsymbol{\mu}_2\ge\mathbf{0}$. Note that the choice of the parameter $\boldsymbol{\mu}$ in (4.27) is critical. A wrong choice of this parameter could lead to either instability and the control gains shooting up to infinity or to poor accuracy (Plestan *et al.*, 2010). Here, we choose $\boldsymbol{\mu}$ as a time-varying parameter function according to (Plestan *et al.*, 2010). Therefore $\boldsymbol{\mu}$ is given by

$$\boldsymbol{\mu}(t) = 4\,\boldsymbol{\alpha}(t)\,T_e\,, \tag{4.28}$$

where T_e is the sampling time for the controller.

An important remark on (4.27) is that the gain adaptation law does not need any knowledge of the upper bound of the external perturbations $\boldsymbol{\kappa}$. Moreover, the gains $\boldsymbol{\alpha}$ and $\boldsymbol{\beta}$ are not chosen according to a worst case uncertainty, but rather they are increased only when necessary. This further reduces the chattering in the ASTC.

Figure 4.2: Physical quadrotor model constructed in SimMechanics.

4.3.2 Feedforward Control

The feedforward component $\bar{\boldsymbol{u}}_{ff}$ based on the dynamic inversion of the nominal model from (4.23) is the wrench that needs to be applied to the nominal model of the UAV to track a reference trajectory, in the absence of initial error. The $\bar{\boldsymbol{u}}_{ff}$ part of the control wrench decreases the magnitude of sliding mode control $\bar{\boldsymbol{u}}_{sm}$, thus helping in reducing the gains of the ASTC and hence attenuates chattering. The expression of $\bar{\boldsymbol{u}}_{ff}$ is obtained by dynamic inversion of (4.19) as

$$\bar{\boldsymbol{u}}_{ff} = \boldsymbol{b}(\boldsymbol{z})^{-1}\left(\begin{bmatrix} \ddddot{x}_d \\ \ddddot{y}_d \\ \ddddot{z}_d \\ \ddot{\psi}_d \end{bmatrix} - \boldsymbol{a}(\boldsymbol{z})\right). \tag{4.29}$$

Figure 4.1 shows the control scheme architecture of the developed controller.

4.4 Physical Simulations

The quadrotor model (4.19), reformulated with the change of coordinates in (4.16), and the capability of the developed adaptive super twisting controller defined by (4.26) and (4.27) are extensively verified by means of physical simulations. We have built a quadrotor system model in SimMechanics[1] using joints, constraints and force elements. SimMechanics formulates and solves the equations of motion for the complete 3D mechanical multibody system and is interfaced with the Matlab/Simulink environment for rapid control design and implementation.

[1]http://www.mathworks.com/products/simmechanics/

Parameter	Description	Value	Unit
m	mass of the UAV	2.6	Kg
g	gravity acceleration	9.81	m/s^2
I_{xx}	inertia along X-axis	0.0488	$Kg.m^2$
I_{yy}	inertia along Y-axis	0.0488	$Kg.m^2$
I_{zz}	inertia along Z-axis	0.0956	$Kg.m^2$
b	lift coefficient	$5.42 * 10^{-5}$	N/Ω^2
d	drag coefficient	$1.1 * 10^{-6}$	Nm/Ω^2
l	arm length	0.215	m
$[\lambda_{x_1}, \lambda_{y_1}, \lambda_{z_1}]$	position error gain	[15, 15, 15]	-
$[\lambda_{x_2}, \lambda_{y_2}, \lambda_{z_2}]$	velocity error gain	[11, 11, 11]	-
$[\lambda_{x_3}, \lambda_{y_3}, \lambda_{z_3}]$	acceleration error gain	[6, 6, 6]	-
λ_{ψ_1}	yaw error gain	1	-
$\boldsymbol{\omega_\alpha}$	positive constant	$diag[200, 200, 200, 20]$	-
$\boldsymbol{\gamma}$	positive constant	$diag[0.8, 0.8, 0.8, 0.8]$	-
$\boldsymbol{\alpha}_m$	minimum positive constant	0.1	-
$\boldsymbol{\eta}$	positive constant	$diag[0.1, 0.1, 0.1, 0.1]$	-
ε	positive constant	$diag[1, 1, 1, 1]$	-
T_e	sampling time	0.001	s

Table 4.1: Experimental parameters

Our aim in this simulation is (i) to prove the robustness of the developed ASTC, (ii) to demonstrate the ability to perform aggressive trajectory tracking maneuvers and (iii) to compare it with standard STC. In the rest of this section we provide a brief description of the experimental setup (Sec. 4.4.1), we show and discuss simulation results of ASTC during aggressive maneuver trajectory tracking (Sec. 4.4.2) and we compare in detail the ASTC with the standard STC (Sec. 4.4.3).

4.4.1 Experimental Setup

The physical quadrotor model in SimMechanics, shown in Fig. 4.2, is designed using the parameters of a real quadrotor with total mass $m = 2.6Kg$ and inertial parameters $\boldsymbol{I}_B = \begin{bmatrix} Ixx & Iyy & Izz \end{bmatrix}^T = \begin{bmatrix} 0.0488 & 0.0488 & 0.0956 \end{bmatrix}^T Kg \cdot m^2$. Note that in the control law these parameters will be considered uncertain. The other system parameters, the lift coefficient b, the drag coefficient d and the arm length l, are considered to be known without uncertainty. Table. 4.1 lists all the experimental parameters used. The system state, namely the position $\boldsymbol{p} = \begin{bmatrix} x & y & z \end{bmatrix}^T$, linear velocity $\dot{\boldsymbol{p}} = \begin{bmatrix} \dot{x} & \dot{y} & \dot{z} \end{bmatrix}^T$, acceleration $\ddot{\boldsymbol{p}} = \begin{bmatrix} \ddot{x} & \ddot{y} & \ddot{z} \end{bmatrix}^T$ in $\mathcal{F}_W$ and the angular velocity $\boldsymbol{\omega_B} = \begin{bmatrix} p & q & r \end{bmatrix}^T$ in $\mathcal{F}_B$ are provided to the controller as noisy measurements, with an additional gaussian noise to resemble

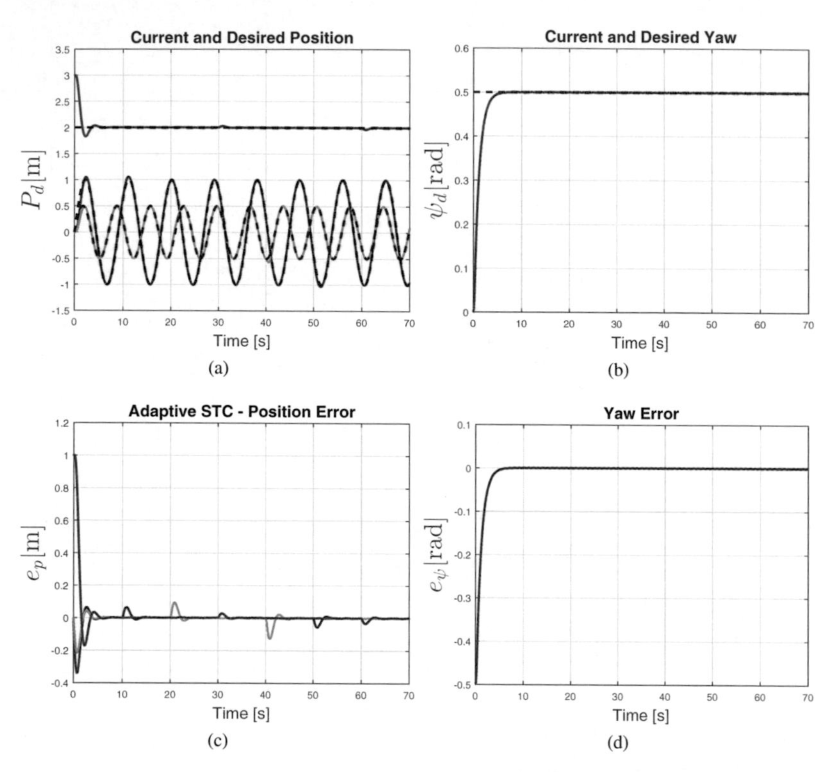

Figure 4.3: Results of robust trajectory tracking for position $\boldsymbol{p}$ and yaw ψ. 4.3(a): Desired (dashed black line) and current (solid line) position $\boldsymbol{p}_d$ in x (red), y (green) and z (blue). 4.3(b): Desired (dashed line) and current (solid line) yaw (red) ψ_d. 4.3(c–d): behavior of the position/orientation tracking errors $(\boldsymbol{e}_p, e_\psi)$.

realistic measurements from an external tracking system and an onboard IMU.

4.4.2 Robustness of ASTC

The desired quadrotor trajectory $\boldsymbol{p}_d$ provided as reference to the controller is a sinusoid along the X and Y axes. The highly aggressive nature of the trajectory is highlighted by the roll and pitch angles during the tracking that reaches up to $\pm 20°$. In order to highlight the robust nature of the controller, the initial position error is set to $\boldsymbol{p}_e = \begin{bmatrix} 0 & 0 & 1 \end{bmatrix}^T m$. Additionally, during the execution of the trajectory, the quadrotor is subjected to high force disturbance in all the principal axes ($f_{\text{ext}_x} = 2N$, $f_{\text{ext}_y} = 3N$ and $f_{\text{ext}_z} = 1N$) which are applied and removed at different time instants, as shown in Fig. 4.4(a). Furthermore, a parameter uncertainty of 10% is included in the controller for the mass m and inertial matrix $\boldsymbol{I}_B$. Therefore this simulation aims to prove the robustness, asymptotic trajectory tracking and stability performance of the controller in presence of initial error, noisy system state, parameter uncertainty and external disturbance.

Figure 4.3(a) shows the desired position $\boldsymbol{p}_d$ and the current position $\boldsymbol{p}$. Figure 4.3(b) shows the desired yaw ψ_d and the current yaw ψ along with the yaw error e_ψ in Fig. 4.3(d). As seen from tracking error $\boldsymbol{e}_p$ in Fig. 4.3(c), the controller shows asymptotic stability even when many nonidealities are are introduced in the model.

Figure 4.4(a) displays the external force disturbance $\boldsymbol{f}_{ext}$ applied on the quadrotor in all the principal axes. The sliding variable $\boldsymbol{\sigma}$, shown in Fig. 4.4(b), $\boldsymbol{\sigma}$ varies with high frequency because of the noise affecting the system state. Figure 4.4(c) shows the adaptation of the $\boldsymbol{\alpha}$ gain given by (4.27). Comparing Fig. 4.4(a) and Fig. 4.4(c), it is possible to notice the spikes in the α gains, due to their adaptation when the disturbance forces are applied or removed. A similar behavior can be observed in the nominal feedforward input computed using (4.29) and shown in Fig. 4.4(d). The control input $\boldsymbol{u}$ and the gain adaptation of $\boldsymbol{\alpha}$ are discussed in detail in Sec. 4.4.3.

4.4.3 Comparison of ASTC and STC

The same physical simulation described in Sec. 4.4.2 is performed also for the standard version of the super twisting controller (STC). Figure 4.5 shows thrust ρ, roll torque τ_x, pitch torque τ_y and yaw torque τ_z computed in the two simulations. It is clear from Fig. 4.5(b) that the control inputs computed by the standard STC are affected by continuous chattering, whereas Fig. 4.5(a) shows that the chattering is substantially reduced and is only present when the gains are adapting to high values to counterbalance the external force disturbance $\boldsymbol{f}_{\text{ext}}$.

Figure 4.6 shows the position error $\boldsymbol{e_p}$ of the ASTC and STC. Clearly, the chattering on the control inputs reflects in a noisy tracking of the desired trajectory (Fig. 4.6(b)), while the ASTC controller shows a smoother behavior (Fig. 4.6(a)). The big difference between the ASTC and standard STC is due to the adaptation of the $\boldsymbol{\alpha}$ gain: while

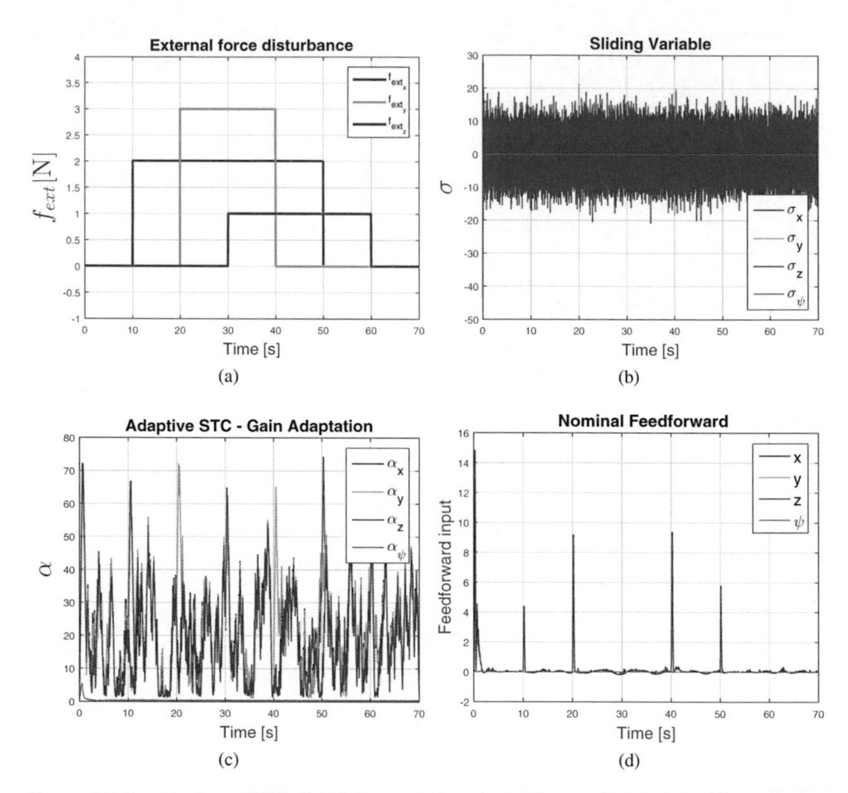

Figure 4.4: Results from ASTC. 4.4(a): External disturbance $\boldsymbol{f}_{ext}$ applied on the quadrotor in f_{ext_x} (red), f_{ext_y} (green) and f_{ext_z} (blue). 4.4(b): Sliding variable $\boldsymbol{\sigma}$ in σ_x (red), σ_y (green), σ_z (blue) and σ_ψ (magenta). 4.4(c): Adaptive $\boldsymbol{\alpha}$ gain of ASTC in α_x (red), α_y (green), α_z (blue) and α_ψ (magenta). 4.4(d): Nominal feedforward proposed in ASTC as ff_x (red), ff_y (green), ff_z (blue) and ff_ψ (magenta)

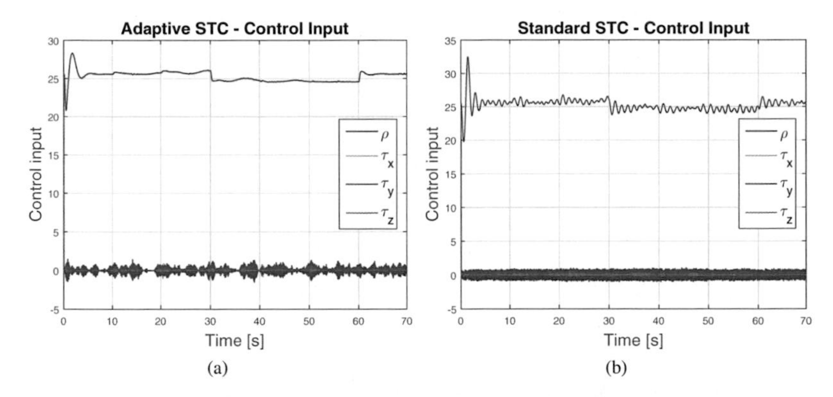

Figure 4.5: Results of the thrust and torque control inputs that are given to the quadrotor. 4.5(a): thrust ρ (red), roll torque τ_x (green), pitch torque τ_y (blue) and yaw torque τ_z (magenta) with ASTC. 4.5(b): thrust ρ (red), roll torque τ_x (green), pitch torque τ_y (blue) and yaw torque τ_z (magenta) with standard STC.

the gains of the STC are constant fixed to $\left(\begin{bmatrix}\alpha_x & \alpha_y & \alpha_z & \alpha_\psi\end{bmatrix} = \begin{bmatrix}20 & 20 & 20 & 2\end{bmatrix}\right)$, the gains of the ASTC are able to vary as shown in Fig. 4.4(c).

4.5 Discussions and Possible Extensions

In this chapter, we have discussed a robust UAV controller and its need in physical interaction and manipulation tasks. To achieve this objective, the problem of trajectory tracking was considered with a quadrotor UAV in presence of uncertainties, external wrenches and noise on the measurements. Therefore

1. it is implemented a robust controller based on a super twisting architecture with adaptive gains (Sec.4.3);
2. the controller is extended to include a feedforward dynamic inversion of the nominal model (Sec.4.3.2);
3. the controller has the advantage of (i) no requirement of knowledge regarding the upper bound of the perturbations (Sec.4.2.3) and (ii) chattering is limited (Sec.4.3.1);
4. it is compared with the standard super twisting to prove the enhancement in the performance (Sec.4.4.3).

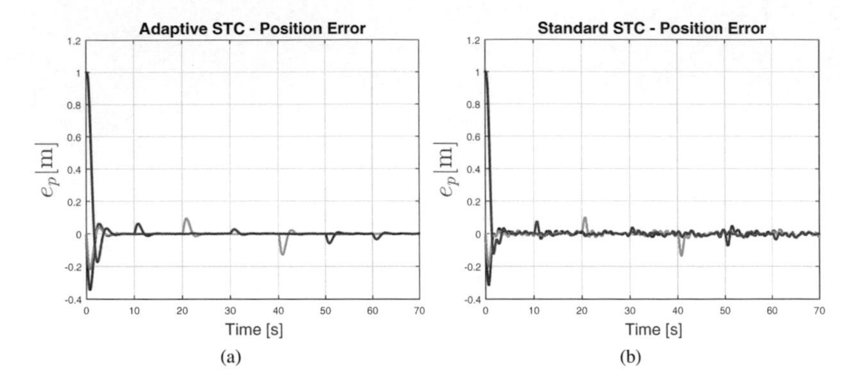

(a) (b)

Figure 4.6: Results of the position tracking error $\boldsymbol{e_p}$ for aggressive maneuvers. 4.6(a): e_x (red), e_y (green), e_z (blue) using ASTC. 4.6(b): e_x (red), e_y (green), e_z (blue) using standard STC.

This adaptive super twisting controller has been proved to be robust and stable. The chattering effect which had been a problem in sliding mode based controllers has been brought to its minimal thanks to adaptive gains. The availability of system states for the higher order dynamic model could be tricky, but it can be estimated from the dynamic model.

Chapter 5

Fully-actuated Hexarotor Aerial Vehicle with Tilted Propellers

The underactuation is a well known characteristic feature of quadrotor UAVs. Initially such UAVs were used for aerial surveillance and mapping applications where trajectory tracking was the main objective. Underactuation was never considered as a serious problem until UAV interaction and manipulation became one of the directions within aerial robotics. Manipulation causes external forces and torques that are generated through interaction with the environment. From a dynamic control point of view, these forces and torques are external perturbations that affect the UAV stability due to the underactuated dynamics of quadrotors, which does not allow to command roll or pitch torque independently with respect to the position. As discussed earlier in Chapter. 3, the underactuation also causes safety issue along with stability when it comes to the application of human-UAV physical interaction. Many researchers started to address the underactuation issue by designing UAV models mostly with complex tilting mechanisms whose final outcome had complex dynamics not fit for precise dynamic modeling and control needed for manipulation and interaction.

So the question still remains as, *How to solve the underactuation problem in UAVs? Is the complex tilting mechanisms proposed earlier the only viable solution? How to design a fully actuated UAV that has simple design parameters and dynamics? How to design a controller for such a UAV? Will it meet the flight stability requirements to perform aerial manipulation? Can such setup be used for human-UAV physical interaction application too?*

In this chapter, we give our answers to these questions by proposing a novel fully actuated hexarotor UAV platform with tilted propellers. We detail the modeling for this new prototype using Newton-Euler formulation and design a feedback linearization control that has simple control properties. Moreover, we validate its performance in the presence of external force / torques and discuss the limitations that exist from the hardware configuration.

The discussion presented in this chapter is based upon the work that I have done under the supervision of Dr. Antonio Franchi during the initial development of the concept phase and Dr. Paolo Stegagno during the experimental phase. The Sec. 5.4.2 on the optimization is done in collaboration with Markus Ryll. Note also that this work has been partly published in Rajappa *et al.* (2015).

5.1 Introduction

As mentioned in the earlier chapters, it has become very popular to utilize aerial vehicles for manipulation and interaction task in environments unsafe and not reachable for humans. As highlighted in Sec. 1.1, the possibility to use UAVs for various civilian tasks such as search and rescue operation, exploration, surveillance, cooperative swarm tasks or transportation are all increasing and has been the main research subject with growing interest in the last decade with many industrial collaborations. New UAV designs, control techniques, sensor utilization and state estimation methodologies are pouring within the UAV community because of the mobile manipulation tasks. Aerial physical interaction, as detailed in Chapter. 3, is starting to make ground and impact because of its safe UAV operation methodology. More research groups are working on direct contact (Marconi *et al.*, 2011; Gioioso *et al.*, 2014b), simple grasping/manipulation tasks (Pounds *et al.*, 2011; Lindsey *et al.*, 2011) and has moved forward to multiple collaborative interactive UAVs (Keemink *et al.*, 2012; Mellinger *et al.*, 2010; Gioioso *et al.*, 2014a).

Among the many challenges faced by typical UAVs, such as little flight time, limited payload capacity, uncertainties in outside environment etc., an important one is the underactuation, i.e., the inability to exert forces in some directions of the body frame. Quadrotors have been used as the main platform for applications as well as research, though they are also underactuated, i.e., they cannot exert any force parallel to the plane perpendicular to their vertical direction in body frame. This is why a quadrotor needs to roll and pitch to accelerate in any direction different from a pure vertical one.

But when it comes to physical interaction, underactuation might become a serious problem for the capabilities and overall stabilization of the aerial vehicle. As the application complexity is going higher, major breakthroughs and advancements in innovative mechanical designs, actuation concepts, micro-electro mechanical systems, sensor technology and power capacity is always envisioned.

5.1.1 Related Works

Several possibilities have been proposed in the past literature spanning different concepts: ducted-fan designs (Naldi *et al.*, 2010), tilt-wing mechanisms (Oner *et al.*, 2008), or tilt-rotor actuations (Kendoul *et al.*, 2006; Sanchez *et al.*, 2008). The concept of tilt-rotor architecture has been much explored to increase flight time (Flores *et al.*, 2011) but not for the improvement of the underactuation problem. In (Salazar *et al.*, 2008) the underactuation was addressed by four additional rotors at the end of each frame in lateral position. But the position of the rotor increased the complexity of controllability because of the air flow between the vertical and the lateral rotors, resulting in non-linear dynamics along with the increase in payload.

A quadrotor design was proposed in (Ryll *et al.*, 2012) with tilted propellers by 4 additional actuators included for the tilting thereby creating the possibility to regulate independently the 6 DoFs of the platform. Though underactuation problem was solved by this design, the need of tilting the propellers in order to resist to any external wrench makes it tough for the aerial manipulation task, where forces shall be exerted instantaneously to resist to unexpected external wrenches. Additionally, the use of servomotors for tilting the propellers makes the overall model challenging to control in real scenarios involving physical interaction. Another tilting mechanism is recently proposed by Odelga *et al.* (2016) through the addition for extra arms and servo motors for tilting the propellers. In Brescianini and D'Andrea (2016), an omni-directional aerial vehicle which is fully actuated and could rotate about in any desired angle is proposed.

In (Voyles and Jiang, 2012), a hexarotor with the propellers rotated by the same angle-magnitude about one axis was suggested. Our approach constitutes a generalization of (Voyles and Jiang, 2012), as we present a more general tilt design. Furthermore, w.r.t. (Voyles and Jiang, 2012) we present a new control law for 6 DoFs trajectory tracking, a methodology to optimize the fixed tilting angles for each propeller depending on the task in exam and an improved mechanical design where all the propellers lie in the same plane.

5.1.2 Methodologies

Taking inspiration from all the related work,

1. it is proposed a novel hexarotor with tilted propeller design, where each rotor is fixedly mounted in a configuration that is rotated about two possible axes. The main objective of this work is full controllability of the UAV's position and orientation by means of tilted propellers, thereby making it completely actuated. The full actuation comes with the acceptable cost of a slightly more complex mechanical design;

2. it is discussed in detail and derived the dynamic model of the proposed tilted propeller hexarotor mechanism;

3. it is devised and developed the closed-loop controller for the hexarotor which is able to asymptotically track an arbitrary desired trajectory for the position and orientation in 3-dimensional free space;

4. it is optimized the propeller tilt angles depending on the application/trajectory to reduce the overall control effort;

5. it is developed the first feasible working model prototype design for the proposed hexarotor with tilt mechanism adaptors;

6. it is validated the hexarotor model and its theoretical concepts both through simulation and experimentally;

5.2 Design and Modeling

A standard hexarotor possesses six propellers that are all rotating about six parallel axes. Even though this choice increases redundancy and payload, such configuration has an underactuated dynamics similar to a standard quadrotor. In fact, the six propellers create an input force that is always parallel to that axis, no matter the values of the six rotational speeds. In this case a change of the direction of the input force in world frame can only be obtained by reorienting the whole vehicle. As a consequence, the output trajectory can only be defined by a 4-dimensional output, namely the center of mass (CoM) 3D position plus the yaw angle, despite the presence of 6 control inputs. In fact in (Mistler *et al.*, 2001) it has been proven that such kind of systems are exactly linearizable with a dynamic feedback using as linearizing output, i.e. the CoM position and the yaw angle. Feedback linearizability also implies differential flatness of the system taking as flat output the linearizing one (De Luca and Oriolo, 2002). The remaining two configuration variables, i.e., the roll and pitch angles, cannot be chosen at will, since they are being determined by the desired trajectory of the CoM, the yaw angle, and their derivatives.

On the converse, the goal of the hexarotor modeling approach presented here is to exploit at best the six available inputs, thus resulting in a system that is fully actuated, i.e., linear and angular accelerations can be set independently acting on the six inputs. In order to obtain full actuation, we remove the constraint for the propellers to rotate about six parallel axes, so that a force in any direction can be generated regardless of the vehicle orientation. Thanks to full actuation, this hexarotor can track 6-DoFs trajectories comprising both the CoM position and, independently, the vehicle orientation described, e.g., by roll, pitch, and yaw, or by a rotation matrix.

Even though a reallocation and reorientation of the six propellers allows for more design flexibility it also increases the number of design parameters thus increasing the design complexity. In order to find a good compromise between full actuation and low number of model parameters, we decide to add the following constraint on the parameters:

- the CoM and the six propeller centers are coplanar, like in a standard hexarotor;

This design choice simplify the design complexity while still allowing a full spectrum of actuation capabilities, as will be shown later in this chapter.

5.2.1 Static System Description

The world inertial frame be denoted with $\mathcal{F}_W : \{\boldsymbol{O}_W, \vec{\boldsymbol{X}}_W, \vec{\boldsymbol{Y}}_W, \vec{\boldsymbol{Z}}_W\}$ and the body frame attached to the hexarotor frame be denoted with $\mathcal{F}_{B_h} : \{\boldsymbol{O}_{B_h}, \vec{\boldsymbol{X}}_{B_h}, \vec{\boldsymbol{Y}}_{B_h}, \vec{\boldsymbol{Z}}_{B_h}\}$, where $\boldsymbol{O}_{B_h}$

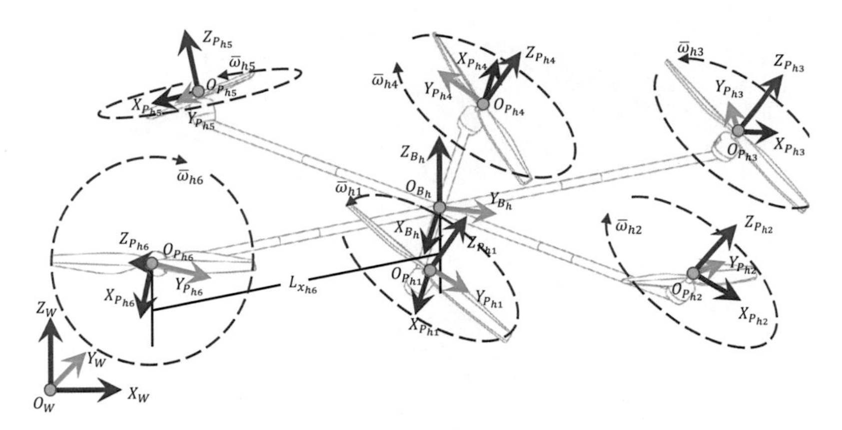

Figure 5.1: Schematic representation of the hexarotor described in this paper.

coincides with the hexarotor CoM. Let the frame associated with the i-th propeller be defined as $\mathcal{F}_{P_{hi}} : \{\boldsymbol{O}_{P_{hi}}, \vec{\boldsymbol{X}}_{P_{hi}}, \vec{\boldsymbol{Y}}_{P_{hi}}, \vec{\boldsymbol{Z}}_{P_{hi}}\}$, where $i = 1\ldots6$. The origin $\boldsymbol{O}_{P_{hi}}$ coincides with the center of spinning and the CoM of the i-th propeller, the axes $\vec{\boldsymbol{X}}_{P_{hi}}$ and $\vec{\boldsymbol{Y}}_{P_{hi}}$ define the rotation plane of the propeller, and $\vec{\boldsymbol{Z}}_{P_{hi}}$ is the axis about which the propeller spins and coincides with the direction of the generated thrust force. The propeller frame $\mathcal{F}_{P_{hi}}$ is rigidly attached to the hexarotor frame, rather than to the propeller, which spins about $\vec{\boldsymbol{Z}}_{P_{hi}}$. In fact, only the direction of the force and torque exerted by the propeller are relevant to our problem. The actual spinning angle of each propeller is not important for the motion, as it will be explained in Sec. 5.2.2.

We shall denote simply by $\boldsymbol{p}_h \in \mathbb{R}^3$ the position of $\boldsymbol{O}_{B_h}$ in $\mathcal{F}_W$, and by ${}^B\boldsymbol{p}_{hi} \in \mathbb{R}^3$ the position of $\boldsymbol{O}_{P_{hi}}$ in $\mathcal{F}_{B_h}$, with $i = 1\ldots6$. In order to have the 6 propellers centers lying on the $\boldsymbol{X}_{B_h}\boldsymbol{Y}_{B_h}$ plane we set:

$$
{}^B\boldsymbol{p}_{hi} = \boldsymbol{R}_Z(\lambda_{hi}) \begin{bmatrix} L_{x_{hi}} \\ 0 \\ 0 \end{bmatrix}, \quad \forall i = 1\ldots6 \tag{5.1}
$$

where $\boldsymbol{R}_Z(\cdot)$ is the canonical rotation matrix about a Z-axis, $L_{x_{hi}} > 0$ is the distance between $\boldsymbol{O}_{P_{hi}}$ and $\boldsymbol{O}_{B_h}$, and λ_{hi} is the angular direction of the segment $\overline{\boldsymbol{O}_{B_h}\boldsymbol{O}_{P_{hi}}}$ on the $\boldsymbol{X}_{B_h}\boldsymbol{Y}_{B_h}$ plane. The Fig. 5.1 show the schematics of different frame references in the tilted propeller hexarotor.

The parameters λ_{hi} and $L_{x_{hi}}$ should be chosen depending on the strength and length of propellers, size and shape of the hexarotor, payload needs, etc.. For example in Sec. 5.5 we shall choose $L_{x_{hi}} = 0.4$ m and $\lambda_{hi} = (i-1)\dfrac{\pi}{3}$.

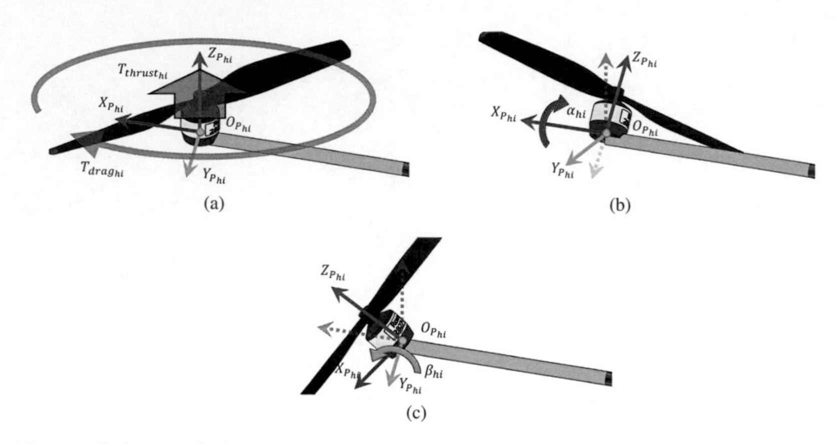

Figure 5.2: (a): terminal part of the i-th hexarotor arm showing the body frame $\mathcal{F}_{P_{hi}}$ and the generated thrust $T_{\text{thrust}_{hi}}$ and drag $T_{\text{drag}_{hi}}$; (b) and (c): Visualization of the possible reorientation of the propeller around $X_{P_{hi}}$ (case (b)) and $Y_{P_{hi}}$ (case (c)). The angle of reorientation is denoted with α_{hi} in (b) and β_{hi} in (c)

Let the rotation matrix ${}^{W}\boldsymbol{R}_{B_h} \in SO(3)$ represent the orientation of $\mathcal{F}_{B_h}$ w.r.t. $\mathcal{F}_W$ and ${}^{B}\boldsymbol{R}_{P_{hi}} \in SO(3)$ represent the orientation of $\mathcal{F}_{P_{hi}}$ w.r.t. $\mathcal{F}_{B_h}$, for $i = 1 \dots 6$. In order to obtain a minimal parameterization of the propeller orientation we decompose each ${}^{B}\boldsymbol{R}_{P_{hi}}$ in three consecutive rotations

$$ {}^{B}\boldsymbol{R}_{P_{hi}} = \boldsymbol{R}_Z(\lambda_{hi})\boldsymbol{R}_X(\alpha_{hi})\boldsymbol{R}_Y(\beta_{hi}), \quad \forall i = 1 \dots 6 \tag{5.2} $$

where the angular parameters α_{hi} and β_{hi} represent the tilt angles that uniquely define the rotation plane of the i-th propeller, $\boldsymbol{X}_{P_{hi}}\boldsymbol{Y}_{P_{hi}}$ or, equivalently, the direction of $\vec{\boldsymbol{Z}}_{P_{hi}}$ in $\mathcal{F}_{B_h}$. The angles α_{hi} and β_{hi} have a clear geometrical interpretation, in fact the i-th propeller plane $\boldsymbol{X}_{P_{hi}}\boldsymbol{Y}_{P_{hi}}$ is obtained from $\boldsymbol{X}_{B_h}\boldsymbol{Y}_{B_h}$ by first applying a rotation of α_{hi} about the line $\overline{\boldsymbol{O}_{B_h}\boldsymbol{O}_{P_{hi}}}$ and then a rotation of β_{hi} about $\vec{\boldsymbol{Y}}_{P_{hi}}$, which lies on $\boldsymbol{X}_{B_h}\boldsymbol{Y}_{B_h}$ and is perpendicular to $\overline{\boldsymbol{O}_{B_h}\boldsymbol{O}_{P_{hi}}}$. The α_{hi} and β_{hi} rotation is pictorially represented in Fig. 5.2.

It would make sense to group the tilt angles of the individual propellers for convenience of modeling into 6-tuples such as $\boldsymbol{\alpha}_h = (\alpha_{h1}, \alpha_{h2}, \alpha_{h3}, \alpha_{h4}, \alpha_{h5}, \alpha_{h6})$, $\boldsymbol{\beta}_h = (\beta_{h1}, \beta_{h2}, \beta_{h3}, \beta_{h4}, \beta_{h5}, \beta_{h6})$ and $\boldsymbol{\lambda}_h = (\lambda_{h1}, \lambda_{h2}, \lambda_{h3}, \lambda_{h4}, \lambda_{h5}, \lambda_{h6})$. Similarly the propeller distance from $\boldsymbol{O}_{B_h}$ can also be grouped as $\boldsymbol{L}_{x_h} = (L_{x_{h1}}, L_{x_{h2}}, L_{x_{h3}}, L_{x_{h4}}, L_{x_{h5}}, L_{x_{h6}})$.

Since it is a tilted propeller setup, it is considered the case in which λ_{hi}, $L_{x_{hi}}$, α_{hi}, β_{hi}, for $i = 1 \dots 6$, are constant during flight. Nevertheless, we allow α_{hi}, β_{hi} to be changed during a pre-flight setup, in order, e.g., to minimize the sum of the overall control effort for a specific task, as shown in Sec. 5.4.2.

5.2.2 Equations of Motion

Utilizing the standard Newton-Euler approach for dynamic systems, it is possible to derive the complete dynamic equations of the tilted propeller hexarotor by considering the forces and torques that are generated by each propeller rotation along with the significant gyroscopic and inertial effects. The following standard[1] assumptions are considered here:

- $\boldsymbol{O}_{B_h}$ coincides with the CoM of the hexarotor;
- $\boldsymbol{O}_{P_{hi}}$ coincides with the CoM of the i-th propeller;
- the motors actuating the six propellers implement a fast high-gain local controller which is able to impose a desired spinning speed with negligible transient, thus allowing to consider the spinning rates as (virtual) control inputs in place of the motor torques;
- gyroscopic and inertial effects due to the propellers and the motors are considered as second-order disturbances to be rejected by the feedback nature of the controller;
- the tilted propellers might cause additional turbulences due to the possible intersection of the airflows. These turbulences are considered as negligible as the possible intersection of the airflows happens not close to the propellers. In fact, tilt configurations have been already proven to be feasible in reality as mentioned in Sec. 5.1.1.

The practicability of these assumptions with the proposed controller on the dynamic model is validated through simulations (see Section 5.5) and experiments (see Section 5.7) which includes the aforementioned unmodeled effects. For ease of presentation and the controller design, in the following modeling we shall express the translational dynamics in $\mathcal{F}_W$ where as the rotational dynamics is expressed in $\mathcal{F}_{B_h}$.

Rotational dynamics

Here in the rotational dynamics, we are concerned with the torques that are generated by each propeller. Since the propellers are tilted in this case, the torques are generated with respect to all the three axes unlike the quadrotor (where torques generated only w.r.t. Z-axis pf propeller). Let us denote with $\boldsymbol{\omega}_{B_h} \in \mathbb{R}^3$ the angular velocity of $\mathcal{F}_{B_h}$, with respect to $\mathcal{F}_W$, expressed in $\mathcal{F}_{B_h}$. Then the rotational dynamics through the standard Newton-Euler formulation is

$$\boldsymbol{I}_{B_h}\dot{\boldsymbol{\omega}}_{B_h} = -\boldsymbol{\omega}_{B_h} \times \boldsymbol{I}_{B_h}\boldsymbol{\omega}_{B_h} + \boldsymbol{\tau}_h + \boldsymbol{\tau}_{h_{\text{ext}}}, \tag{5.3}$$

[1]Similar assumptions have been used, e.g., in (Flores *et al.*, 2011; Salazar *et al.*, 2008; Ryll *et al.*, 2012)

where $\boldsymbol{I}_{B_h}$ is the hexarotor body inertia matrix, $\boldsymbol{\tau}_{h_{\text{ext}}}$ accounts for external disturbances and unmodeled effects, and $\boldsymbol{\tau}_h$ is the input torque which is further decomposed in

$$\boldsymbol{\tau}_h = \boldsymbol{\tau}_{h_{\text{thrust}}} + \boldsymbol{\tau}_{h_{\text{drag}}}, \tag{5.4}$$

where $\boldsymbol{\tau}_{h_{\text{thrust}}}$ is the torque generated by the six tilted propeller thrusts and $\boldsymbol{\tau}_{h_{\text{drag}}}$ is due to the six propeller drags. The two individual components of (5.4) are discussed in detail below.

Torque due to thrusts ($\boldsymbol{\tau}_{h_{\text{thrust}}}$): The i-th propeller rotation creates a force vector applied at $\boldsymbol{O}_{P_{hi}}$ and directed along $\vec{\boldsymbol{Z}}_{P_{hi}}$, which is expressed in $\mathcal{F}_{P_{hi}}$ by

$$\boldsymbol{T}_{\text{thrust}_{hi}} = \begin{bmatrix} 0 & 0 & k_f \bar{\omega}_{hi}^2 \end{bmatrix}^T \tag{5.5}$$

where $k_f > 0$ is a constant thrust coefficient and $\bar{\omega}_{hi}$ is the spinning velocity of the i-th propeller. This thrust torque that is generated at each propeller in its own frame $\mathcal{F}_{P_{hi}}$ can be transferred to $\mathcal{F}_{B_h}$ by

$$\boldsymbol{\tau}_{h_{\text{thrust}}} = \sum_{i=1}^{6} \left({}^B\boldsymbol{p}_{hi} \times {}^B\boldsymbol{R}_{P_{hi}} \boldsymbol{T}_{\text{thrust}_{hi}} \right). \tag{5.6}$$

Torque due to drag ($\boldsymbol{\tau}_{h_{\text{drag}}}$): The drag moment generated by the i-th propeller acts in the opposite direction of the propeller angular velocity and is expressed in $\mathcal{F}_{P_{hi}}$ by

$$\boldsymbol{T}_{\text{drag}_{hi}} = \begin{bmatrix} 0 & 0 & (-1)^i k_m \bar{\omega}_{hi}^2 \end{bmatrix}^T, \tag{5.7}$$

where $k_m > 0$ is the propeller drag coefficient. Note that the drag component also acts w.r.t. $\vec{\boldsymbol{Z}}_{P_{hi}}$. The factor $(-1)^i$ is used since half of the propellers rotate clockwise and the other half rotates counter-clockwise. This is done in order to have an automatic counterbalance of the drag torques at hovering. The drag torque due to the six propellers expressed in $\mathcal{F}_{B_h}$ is then

$$\boldsymbol{\tau}_{h_{\text{drag}}} = \sum_{i=1}^{6} {}^B\boldsymbol{R}_{P_{hi}} \boldsymbol{T}_{\text{drag}_{hi}}. \tag{5.8}$$

Now combining together the generated torque by thrust (5.6) and drag (5.8) as the input torque in (5.4), we can write

$$\boldsymbol{\tau}_h = \sum_{i=1}^{6} \left({}^B\boldsymbol{p}_{hi} \times {}^B\boldsymbol{R}_{P_{hi}} \boldsymbol{T}_{\text{thrust}_{hi}} \right) + \sum_{i=1}^{6} {}^B\boldsymbol{R}_{P_{hi}} \boldsymbol{T}_{\text{drag}_{hi}} \tag{5.9}$$

$$\boldsymbol{\tau}_h = \boldsymbol{H}(\boldsymbol{\alpha}_h, \boldsymbol{\beta}_h, \boldsymbol{\lambda}_h, \boldsymbol{L}_{x_h})\boldsymbol{u}_h, \tag{5.10}$$

where $\boldsymbol{H}(\boldsymbol{\alpha}_h, \boldsymbol{\beta}_h, \boldsymbol{\lambda}_h, \boldsymbol{L}_{x_h}) \in \mathbb{R}^{3\times 6}$ is the matrix that relates the input torque $\boldsymbol{\tau}_h$ to the control input

$$\boldsymbol{u}_h = [\bar{\omega}_{h1}^2 \ \bar{\omega}_{h2}^2 \ \bar{\omega}_{h3}^2 \ \bar{\omega}_{h4}^2 \ \bar{\omega}_{h5}^2 \ \bar{\omega}_{h6}^2]^T \in \mathbb{R}^{6\times 1}, \tag{5.11}$$

i.e., the squares of the rotational speeds of each propeller. The individual component of $\boldsymbol{H}(\boldsymbol{\alpha}_h, \boldsymbol{\beta}_h, \boldsymbol{\lambda}_h, \boldsymbol{L}_{x_h})$ can be seen in Appendix. A.1.2 along with its detailed computation.

Translational dynamics

The assumptions earlier on the location of the hexarotor body frame and propeller centers of mass help to express the translational dynamics in $\mathcal{F}_W$, using the standard Newton-Euler formulation, as

$$m_h \ddot{\boldsymbol{p}}_h = m_h \begin{bmatrix} 0 \\ 0 \\ -g \end{bmatrix} + {}^W\boldsymbol{R}_{B_h} \boldsymbol{F}(\boldsymbol{\alpha}_h, \boldsymbol{\beta}_h, \boldsymbol{\lambda}_h)\boldsymbol{u}_h + \boldsymbol{f}_{h_{\text{ext}}} \tag{5.12}$$

where m_h is the hexarotor mass, $\boldsymbol{f}_{h_{\text{ext}}}$ represents external disturbances and unmodeled effects, and $\boldsymbol{F}(\boldsymbol{\alpha}_h, \boldsymbol{\beta}_h, \boldsymbol{\lambda}_h) \in \mathbb{R}^{3\times 6}$ is the matrix that relates $\boldsymbol{u}_h$ with the total force produced by the each tilted propellers which is expressed in body frame, i.e.,

$$\boldsymbol{F}(\boldsymbol{\alpha}_h, \boldsymbol{\beta}_h, \boldsymbol{\lambda}_h)\boldsymbol{u}_h = \sum_{i=1}^{6} {}^B\boldsymbol{R}_{P_{hi}} \boldsymbol{T}_{\text{thrust}_{hi}}. \tag{5.13}$$

Notice that in a standard hexarotor $\alpha_{hi} = \beta_{hi} = 0$, for all $i = 1\ldots 6$ which implies that $\boldsymbol{F}(\boldsymbol{\alpha}_h, \boldsymbol{\beta}_h, \boldsymbol{\lambda}_h)$ has rank equal to one (the total force is always directed on the $\vec{\boldsymbol{Z}}_{B_h}$ axis). Here in the case of tilted propeller hexarotor $\alpha_{hi} \neq 0$ and $\beta_{hi} \neq 0$. This makes sure that the rank of $\boldsymbol{F}(\boldsymbol{\alpha}_h, \boldsymbol{\beta}_h, \boldsymbol{\lambda}_h)$ would be three, which is the minimum required for independently commanding any desired position through its translational dynamics. This is further detailed in Sec. 5.4.1. Furthermore, the individual components of $\boldsymbol{F}(\boldsymbol{\alpha}_h, \boldsymbol{\beta}_h, \boldsymbol{\lambda}_h)$ can be seen in Appendix. A.1.1, which also provides a detailed insight in the computation of translational dynamics.

5.3 Control Design

With the system dynamics clearly articulated, the next biggest step would be to design an appropriate controller that may very well consider the modeling complexity but keeps the control characteristic simple and approachable so that this platform could be easily used for manipulation and interaction, which has its own complex framework. The control

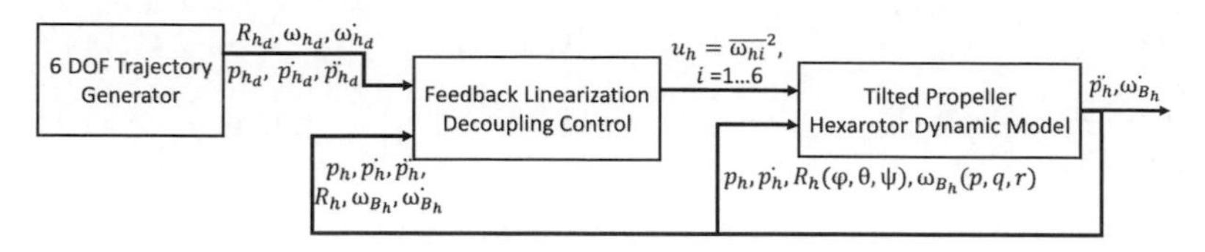

Figure 5.3: Control scheme architecture

problem considered here is an output tracking problem. In particular, the hexarotor is tasked to track, simultaneously, a desired trajectory $\boldsymbol{p}_{h_d}(t)$ with the CoM position $\boldsymbol{p}_h$ and a given orientation $\boldsymbol{R}_{h_d}(t)$ with the body orientation ${}^W\boldsymbol{R}_{B_h}$. The available control inputs are the squares of the six spinning rates of the propellers $\boldsymbol{u}_h$ defined in (5.11).

The dynamical model of the tilted propeller hexarotor derived earlier can be rewritten for the control design as follows,

$$\ddot{\boldsymbol{p}}_h = \begin{bmatrix}0 & 0 & -g\end{bmatrix}^T + \frac{1}{m_h}{}^W\boldsymbol{R}_{B_h}\boldsymbol{F}(\boldsymbol{\alpha}_h,\boldsymbol{\beta}_h,\boldsymbol{\lambda}_h)\boldsymbol{u}_h \tag{5.14}$$

$$\dot{\boldsymbol{\omega}}_{B_h} = -\boldsymbol{I}_{B_h}^{-1}(\boldsymbol{\omega}_{B_h}\times \boldsymbol{I}_{B_h}\boldsymbol{\omega}_{B_h}) + \boldsymbol{I}_{B_h}^{-1}\boldsymbol{H}(\boldsymbol{\alpha}_h,\boldsymbol{\beta}_h,\boldsymbol{\lambda}_h,\boldsymbol{L}_{x_h})\boldsymbol{u}_h \tag{5.15}$$

$${}^W\dot{\boldsymbol{R}}_{B_h} = {}^W\boldsymbol{R}_{B_h}[\boldsymbol{\omega}_{B_h}]_\wedge \tag{5.16}$$

with $[\cdot]_\wedge$ being the hat operator from $\mathbb{R}^3$ to $so(3)$. Here the external forces and torques are neglected considering them as disturbances which are handled by the feedback nature of the controller.

5.3.1 Exact Feedback Linearization and Decoupling Control

The exact feedback linearization and decoupling control (Khalil, 2002) falls under the class of non-linear control technique. The principle behind the functioning of such a controller is primarily through the decoupling of control input and output, such that the decoupling matrix which relates them has full rank (in this hexarotor case, it is 6). The full ranked decoupling matrix later, after the application of appropriate control action, makes sure the desired trajectory is tracked through static feedback. In certain scenarios, if the decoupling matrix is not full ranked, then the system dynamics could be differentiated further to higher-order until this condition is fulfilled through dynamic feedback linearization.

In order to apply a feedback linearization technique we rewrite (5.14)–(5.15) in a matricial form

$$\begin{bmatrix}\ddot{\boldsymbol{p}}_h \\ \dot{\boldsymbol{\omega}}_{B_h}\end{bmatrix} = \boldsymbol{f}_h + \boldsymbol{J}_{R_h}\left[\bar{\boldsymbol{J}}(\boldsymbol{\alpha}_h,\boldsymbol{\beta}_h,\boldsymbol{\lambda}_h,\boldsymbol{L}_{x_h})\right]\boldsymbol{u}_h = \boldsymbol{f}_h + \boldsymbol{J}(\boldsymbol{\alpha}_h,\boldsymbol{\beta}_h,\boldsymbol{\lambda}_h,\boldsymbol{L}_{x_h})\boldsymbol{u}_h \tag{5.17}$$

where $\boldsymbol{f}_h \in \mathbb{R}^6$ is the drift vector due to the gravity and the rotational inertia, $\boldsymbol{J}_{R_h} = \begin{bmatrix} \frac{1}{m_h}{}^W\boldsymbol{R}_{B_h} & \boldsymbol{0} \\ \boldsymbol{0} & \boldsymbol{I}_{B_h}^{-1} \end{bmatrix} \in \mathbb{R}^{6\times 6}$, $\bar{\boldsymbol{J}}(\boldsymbol{\alpha}_h, \boldsymbol{\beta}_h, \boldsymbol{\lambda}_h, \boldsymbol{L}_{x_h}) = \begin{bmatrix} \boldsymbol{F}(\boldsymbol{\alpha}_h, \boldsymbol{\beta}_h, \boldsymbol{\lambda}_h) \\ \boldsymbol{H}(\boldsymbol{\alpha}_h, \boldsymbol{\beta}_h, \boldsymbol{\lambda}_h, \boldsymbol{L}_{x_h}) \end{bmatrix} \in \mathbb{R}^{6\times 6}$. The 6×6 matrix $\boldsymbol{J}(\boldsymbol{\alpha}_h, \boldsymbol{\beta}_h, \boldsymbol{\lambda}_h, \boldsymbol{L}_{x_h})$ is called the *decoupling matrix*[2]. This matrix relates the system dynamics with the control input as mentioned earlier. If $\boldsymbol{J}(\boldsymbol{\alpha}_h, \boldsymbol{\beta}_h, \boldsymbol{\lambda}_h, \boldsymbol{L}_{x_h})$ is invertible then the control input can be chosen as

$$\boldsymbol{u}_h = \boldsymbol{J}^{-1}(\boldsymbol{\alpha}_h, \boldsymbol{\beta}_h, \boldsymbol{\lambda}_h, \boldsymbol{L}_{x_h})(-\boldsymbol{f}_h + \boldsymbol{v}_h) \tag{5.18}$$

where $\boldsymbol{v}_h$ is an additional virtual input injected to take the appropriate control action depending upon the error, thus obtaining

$$\begin{bmatrix} \ddot{\boldsymbol{p}}_h \\ \dot{\boldsymbol{\omega}}_{B_h} \end{bmatrix} = \boldsymbol{v}_h = \begin{bmatrix} \boldsymbol{v}_{h_p} \\ \boldsymbol{v}_{h_R} \end{bmatrix}, \tag{5.19}$$

i.e., the system is exactly linearized via a static feedback. Fig. 5.3 shows the control scheme architecture.

In order to obtain an exponential convergence to $\boldsymbol{0}$ of the position error $\boldsymbol{p}_h - \boldsymbol{p}_{h_d} = \boldsymbol{e}_{h_p}$ one can choose a linear controller

$$\boldsymbol{v}_{h_p} = \ddot{\boldsymbol{p}}_{h_d} - \boldsymbol{K}_{h_{p1}}\dot{\boldsymbol{e}}_{h_p} - \boldsymbol{K}_{h_{p2}}\boldsymbol{e}_{h_p} - \boldsymbol{K}_{h_{p3}}\int_{t_0}^{t} \boldsymbol{e}_{h_p}, \tag{5.20}$$

where the diagonal positive definite gain matrixes $\boldsymbol{K}_{h_{p1}}$, $\boldsymbol{K}_{h_{p2}}$, $\boldsymbol{K}_{h_{p3}}$ define Hurwitz polynomials. This control law compensates for any integral error arising in the translational system dynamics as well.

Now considering the orientation tracking, a popular used parameterization is to resort to Euler angles. However it is well known that they are prone to singularity problems. Keeping this in mind, the controller for the rotational configuration is developed directly on $SO(3)$ and thereby it avoids any singularities that arise in local coordinates, such as Euler angles. Now assuming that $\boldsymbol{R}_{h_d}(t) \in \bar{C}^3$ and $\boldsymbol{\omega}_{h_d} = [\boldsymbol{R}_{h_d}^T \dot{\boldsymbol{R}}_{h_d}]_\vee$, where $[\cdot]_\vee$ represents the inverse (vee) operator from $so(3)$ to $\mathbb{R}^3$, the attitude tracking error $\boldsymbol{e}_{h_R} \in \mathbb{R}^3$ is defined similarly to (Lee *et al.*, 2010) as

$$\boldsymbol{e}_{h_R} = \frac{1}{2}[\boldsymbol{R}_{h_d}^T\, {}^W\boldsymbol{R}_{B_h} - {}^W\boldsymbol{R}_{B_h}^T \boldsymbol{R}_{h_d}]_\vee, \tag{5.21}$$

and the tracking error of the angular velocity $\boldsymbol{e}_{h_\omega} \in \mathbb{R}^3$ is given by

$$\boldsymbol{e}_{h_\omega} = \boldsymbol{\omega}_{B_h} - {}^W\boldsymbol{R}_{B_h}^T \boldsymbol{R}_{h_d} \boldsymbol{\omega}_{h_d}. \tag{5.22}$$

[2] In standard hexarotor the decoupling matrix $\boldsymbol{J}(\boldsymbol{\alpha}_h, \boldsymbol{\beta}_h, \boldsymbol{\lambda}_h, \boldsymbol{L}_{x_h})$ has always rank equal to four, similarly to a quadrotor.

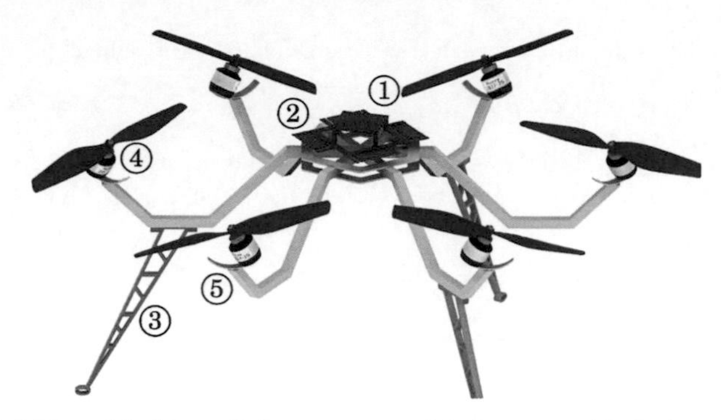

Figure 5.4: CAD model of the preliminary prototype of the hexarotor with tilted propellers. It is composed of: (1) Micro controller, (2) Brushless controller, (3) Lander, (4) Propeller motor, (5) Tilting set-up.

In order to obtain an asymptotic convergence to $\mathbf{0}$ of the rotational error $\boldsymbol{e}_{h_R}$ one can choose the following controller

$$\boldsymbol{v}_{h_R} = \dot{\boldsymbol{\omega}}_{h_d} - \boldsymbol{K}_{h_{R1}}\boldsymbol{e}_{h_\omega} - \boldsymbol{K}_{h_{R2}}\boldsymbol{e}_{h_R} - \boldsymbol{K}_{h_{R3}}\int_{t_0}^{t}\boldsymbol{e}_{h_R} \tag{5.23}$$

where the diagonal positive definite gain matrixes $\boldsymbol{K}_{h_{R1}}, \boldsymbol{K}_{h_{R2}}, \boldsymbol{K}_{h_{R3}}$ define Hurwitz polynomials also in this case.

5.4 A Preliminary Prototype

In this section we present the design of a preliminary prototype obtained instantiating the general model introduced in Section 5.2 in a more particular case. A CAD of the prototype is shown in Fig. 5.4. In order to reduce the complexity and for the sake of symmetry, it has been chosen $\lambda_{hi} = (i-1)\frac{\pi}{3}$ and $L_{x_{hi}} = 0.4$ m $\forall\ i = 1\ldots6$. With this choice, the origin $\boldsymbol{O}_{P_{hi}}$ of each propeller frame is equally spaced with $60°$ between each other from the center of the body frame $\boldsymbol{O}_{B_h}$ to have a symmetric configuration in normal hovering position.

Furthermore, as shown in Fig. 5.5 each propeller is mounted in an arc frame which is free to rotate in $\vec{\boldsymbol{X}}_{P_{hi}}$ and $\vec{\boldsymbol{Y}}_{P_{hi}}$, so that the tilt angle of α_{hi} and β_{hi} can be fixed as desired. The radius of the arc(R_{arc}) is designed equal to the length of the motor (with the propeller attached), so that $\boldsymbol{O}_{P_{hi}}$ always stays at the same location in the $\boldsymbol{X}_{B_h}\boldsymbol{Y}_{B_h}$ plane with only its direction vector $[\boldsymbol{X}_{P_{hi}}\ \boldsymbol{Y}_{P_{hi}}\ \boldsymbol{Z}_{P_{hi}}]^T$ changing according to the α_{hi} and β_{hi} orientation.

The arm in which each propeller set-up is suspended is designed to have a curved architecture with the radius of the curvature, more than the propeller radius ($R_{h_{prop}}$),

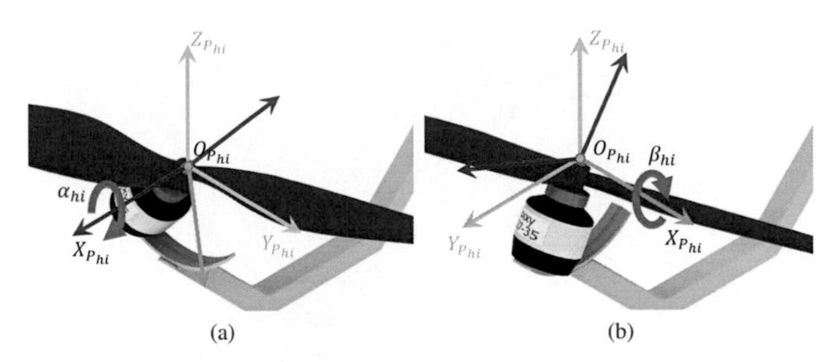

Figure 5.5: (a) and (b): Visualization of the possible reorientation of the propeller around X_{P_i} (case (a)) and Y_{P_i} (case (b)). The angle of reorientation is denoted with α_i in (a) and β_i in (b)

so that independently from the value of α_{hi} and β_{hi} in a certain allowed interval, the propellers never come in contact with the arm during flight. Finally in this preliminary prototype we consider the following constraints

$$\alpha_{h1} = \pm\alpha_{h2} = \pm\alpha_{h3} = \pm\alpha_{h4} = \pm\alpha_{h5} = \pm\alpha_{h6} = \boldsymbol{\alpha}_h \tag{5.24}$$

$$\beta_{h1} = \pm\beta_{h2} = \pm\beta_{h3} = \pm\beta_{h4} = \beta_{h5} = \pm\beta_{h6} = \boldsymbol{\beta}_h. \tag{5.25}$$

This constraint makes sure that symmetric configuration is maintained from the location where individual propeller thrust is generated. Furthermore, this configuration also helps reducing the energy consumption thereby creating an optimum tilt angle. This is further detailed in Sec. 5.4.2.

5.4.1 Discussion on the Invertibility of $J(\alpha_h, \beta_h)$

In this prototype the decoupling matrix depends only on the choice of $\boldsymbol{\alpha}_h$ and $\boldsymbol{\beta}_h$ which are the varying parameters, the rest being constant. As mentioned earlier, the exact feedback linearization control algorithm in this case relies on the invertibility of $\boldsymbol{J}(\boldsymbol{\alpha}_h, \boldsymbol{\beta}_h)$. This implies $\rho_J = rank(\boldsymbol{J}(\boldsymbol{\alpha}_h, \boldsymbol{\beta}_h)) = rank(\boldsymbol{J}_{R_h}\bar{\boldsymbol{J}}(\boldsymbol{\alpha}_h, \boldsymbol{\beta}_h)) \equiv rank(\bar{\boldsymbol{J}}(\boldsymbol{\alpha}_h, \boldsymbol{\beta}_h)) = 6, \forall t > 0$. Here $\boldsymbol{J}_{R_h}$ is a nonsingular square matrix as seen in (5.17) and therefore does not affect $rank(\boldsymbol{J}(\boldsymbol{\alpha}_h, \boldsymbol{\beta}_h))$. Therefore $\bar{\boldsymbol{J}}(\boldsymbol{\alpha}_h, \boldsymbol{\beta}_h)$ is the only rank affecting component. The individual components of this can be seen in Appendix. A.1. Due to the high non linearity of $\bar{\boldsymbol{J}}(\boldsymbol{\alpha}_h, \boldsymbol{\beta}_h)$ sufficient conditions for the invertibility are hard to find. Fig. 5.6 graphically shows the determinant value $\det(\bar{\boldsymbol{J}}(\boldsymbol{\alpha}_h, \boldsymbol{\beta}_h))$ for a particular choice of the pluses and minuses in (5.24) and (5.25). This determinant value being zero indirectly affects the invertibility of $\bar{\boldsymbol{J}}(\boldsymbol{\alpha}, \boldsymbol{\beta})$ which would affect the controllability of the hexarotor. Therefore the objective is to avoid any such singularity causing configuration should be avoided during the prefixing of the propeller tilt-angle.

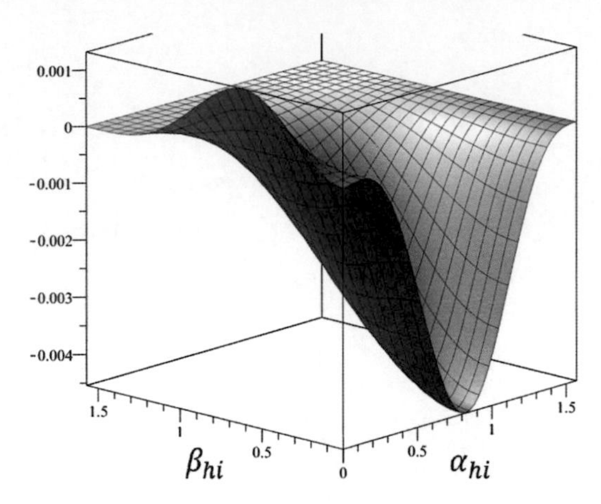

Figure 5.6: Determinant value (z-axis) of $\bar{\boldsymbol{J}}(\boldsymbol{\alpha}_{hi}, \boldsymbol{\beta}_{hi})$ of the presented prototype

5.4.2 Optimization of α_h and β_h

The angles $\boldsymbol{\alpha}_h$ and $\boldsymbol{\beta}_h$ can be adjusted during the pre-flight setup. This gives the possibility to change the angles depending on the needs of a particular trajectory. In this section, we consider this capability to optimize $\boldsymbol{\alpha}_h$ and $\boldsymbol{\beta}_h$ depending on a predefined desired trajectory to reduce the control effort. As a reminder, the main motive is the full controllability in position and orientation. This comes with the cost of a higher control effort. The objective of this section is therefore to reduce this parasitic effect.

The predominant energy consuming parts of the hexarotor are the propeller motors. Minimizing the control effort through the norm of the control output $||\boldsymbol{u}_h||$ by optimizing the particular $\boldsymbol{\alpha}_h$ and $\boldsymbol{\beta}_h$ will as well reduce the energy consumption in flight. To reduce the complexity of the optimization, α_{hi} and β_{hi} shall be changed in a coordinated way as explained before. We decided to use the same $\boldsymbol{\alpha}_h$ and $\boldsymbol{\beta}_h$ respectively for α_{hi} and β_{hi} $\in i = 1 \ldots 6$, but with different signs for the individual joints (see (5.24) and (5.25)). An overview of the compared configurations can be found in table 5.1.

The coordinated variation of α_{hi} and β_{hi} offers two additional advantages: (i) no asymmetries in the hexarotor body and (ii) none or a minimum change of the CoM. Considering these constraints, the optimization problem can be defined as:

$$\min_{\boldsymbol{\alpha}_h, \boldsymbol{\beta}_h} \int_0^{t_f} ||\boldsymbol{u}_h|| \mathrm{d}t \tag{5.26}$$

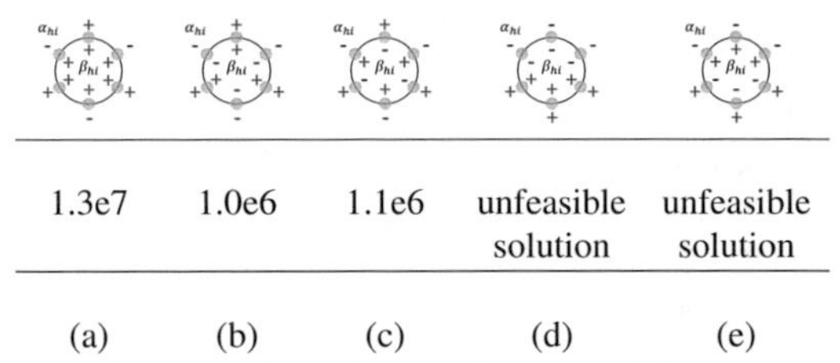

(a)	(b)	(c)	(d)	(e)
1.3e7	1.0e6	1.1e6	unfeasible solution	unfeasible solution

Table 5.1: Stylized tested configuration and results. First row: Different configuration presented. Outside the circle the sign of α_{hi} is indicated, within the circle the sign of β_{hi} is indicated. Second row: Value of the optimized $\int ||\boldsymbol{u}_h||_{min}$. Configuration (b) is the best configuration for the given trajectory. Configurations (d) and (e) are not feasible configurations

Subject to:

$$0 < \boldsymbol{\alpha}_h < \frac{\pi}{2} \tag{5.27}$$

$$0 < \boldsymbol{\beta}_h < \frac{\pi}{2} \tag{5.28}$$

$$0 < \bar{\omega}_{hi} \text{ , for } i \in 1..6 \tag{5.29}$$

Here (5.27) and (5.28) are defining the lower and upper bounds for α_{hi} and β_{hi}, while (5.29) ensures a positive rotation speed $\bar{\omega}_{hi}$ for all propellers. The presented minimization problem is a multi-dimensional constrained nonlinear optimization problem and can be solved using the in-build optimization capabilities of MATLAB by exploiting the *fmincon*-function (Cheon *et al.*, 2013).

To compare (minimal control effort) the different configurations shown in table 5.1, we used the presented optimization technique to find the optimal values $\alpha_h^\star$ and $\beta_h^\star$ and the associated $\int ||\omega_h||_{min}$. As trajectory, a typical flight regime has been chosen, which is presented in Section 5.5.1. The minimum value of the objective function could be found in configuration (b). Therefore all further experiments will be performed by using this configuration: $\boldsymbol{\alpha}_h = \alpha_{h1} = -\alpha_{h2} = \alpha_{h3} = -\alpha_{h4} = \alpha_{h5} = -\alpha_{h6}$ and $\boldsymbol{\beta}_h = \beta_{h1} = -\beta_{h2} = \beta_{h3} = -\beta_{h4} = \beta_{h5} = -\beta_{h6}$.

The optimal angles $\alpha_h^\star$ and $\beta_h^\star$ are highly dependent on the desired trajectory. To visualize the influence we conducted a trajectory, where the hexarotor hovers in place but performs a sinusoidal rotation around θ and ϕ at the same time (see figure 5.7 (a)). The magnitude of the rotation is increased in 6 steps up to 22.5°. $\alpha_h^\star$ and $\beta_h^\star$ are increasing accordingly from almost zero values to $\alpha_h^\star = 0.49$ rad and $\beta_h^\star = 0.33$ rad for the maximum amplitude. Figure 5.8 shows the influence of the optimization itself. For the considered sinusoidal trajectory, the value of the objective function for a wide variety of $\alpha_h^\star$ and $\beta_h^\star$ was calculated. The optimal value is marked by a red circle in Fig. 5.8.

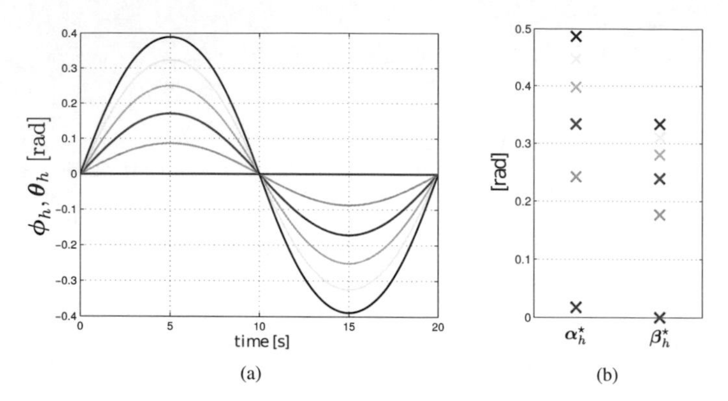

Figure 5.7: (a): Desired sinusoidal trajectories for equal θ_h and ϕ_h. Their amplitude is increased in six steps from 0° (0 rad) to 22.5° (≈ 0.39 rad); all other values remain constant ($= 0$). (b): Optimal values for α_h and β_h corresponding to the six trajectories presented in (a)

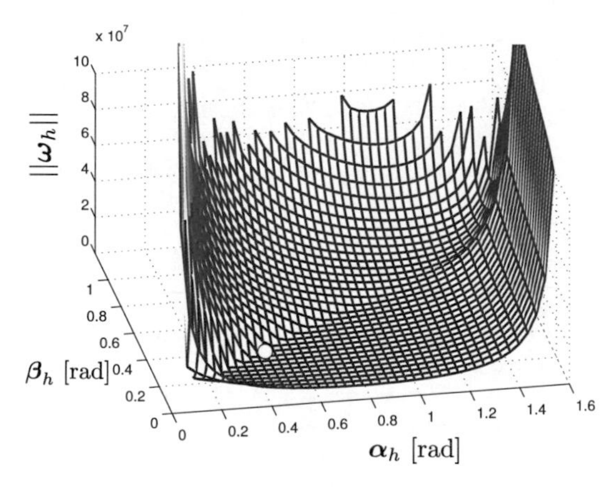

Figure 5.8: Objective function for a given trajectory. Optimum is marked with a red circle

5.5 Simulations and Analysis

Here, the intend is to present two simulations performed on the novel tilted propeller hexarotor. The aim is to prove two important features of the proposed platform: (i) the ability of reorienting while hovering and under the influence of external force/torque disturbances, (ii) the task of 6-DoF (position+orientation) trajectory tracking. Clearly the standard quadrotor cannot perform the above tasks because of the underactuation. Therefore these simulations would prove the full actuation. Given the chosen $\boldsymbol{\alpha}_h$ and $\boldsymbol{\beta}_h$, not all trajectories might be feasible since the negative control outputs $\boldsymbol{u}_h$ could get generated. Therefore, the tilt-angle is carefully selected during the pre-trajectory planning step and not kept in any singularity causing tilt or angles which results in infeasible propeller velocities.

5.5.1 Reorienting while hovering with external disturbance

In the first simulation, we tested a hovering trajectory in which the hexarotor maintains a fixed position $\boldsymbol{p}_h$ but re-orients itself changing at the same time the roll ϕ, pitch θ and yaw ψ angles. This involves hexarotor orienting $-12°$ w.r.t. $\vec{\boldsymbol{X}}_{B_h}$, $12°$ w.r.t. $\vec{\boldsymbol{Y}}_{B_h}$ axis and $15°$ w.r.t. $\vec{\boldsymbol{Z}}_{B_h}$ while still hovering in the position $\boldsymbol{p}_h = [0\ \ 0\ \ 0]^T$. Notice that orienting w.r.t. the 3 principal body axes $\{\vec{\boldsymbol{X}}_{B_h}, \vec{\boldsymbol{Y}}_{B_h}, \vec{\boldsymbol{Z}}_{B_h}\}$ while holding the same position is not feasible in a standard (co-planar) hexarotor UAV. The initial conditions were set to $\boldsymbol{p}_h(t_0) = \boldsymbol{0}$, $\dot{\boldsymbol{p}}_h(t_0) = \boldsymbol{0}$, ${}^W\boldsymbol{R}_{B_h}(t_0) = \boldsymbol{I}_3$ and $\boldsymbol{\omega}_{B_h}(t_0) = \boldsymbol{0}$. The desired trajectory was chosen as $\boldsymbol{p}_{h_d}(t) = 0$ and $\boldsymbol{R}_{h_d}(t) = \boldsymbol{R}_X(\phi(t))\boldsymbol{R}_Y(\theta(t))\boldsymbol{R}_Z(\psi(t))$ with $\phi(t)$, $\theta(t)$, $\psi(t)$ following a smooth profile having as maximum velocity $\dot{\theta}_{max} = 5°/\text{s}$ and maximum acceleration $\ddot{\theta}_{max} = 2.5°/\text{s}^2$. The optimized value of $\alpha_h' = 13.6°$ and $\beta_h' = 10.6°$ obtained from Sec. 5.4.2 has been used. The gains in Equations (5.20) and (5.23) were set to $\boldsymbol{K}_{h_{p1}} = \boldsymbol{K}_{h_{R1}} = 10\boldsymbol{I}_3$, $\boldsymbol{K}_{h_{p2}} = \boldsymbol{K}_{h_{R2}} = 29\boldsymbol{I}_3$ and $\boldsymbol{K}_{h_{p3}} = \boldsymbol{K}_{h_{R3}} = 30\boldsymbol{I}_3$.

Figures 5.9(a–d) show the result of hovering with external force/torque disturbance. As clearly seen in Fig. 5.9(c) a constant external force disturbance ($\boldsymbol{f}_{h_{\text{ext}}} = [4\ \ 2\ \ 1]^T\text{N}$) is applied, along the 3 principal axis $\{\vec{\boldsymbol{X}}_{B_h}, \vec{\boldsymbol{Y}}_{B_h}, \vec{\boldsymbol{Z}}_{B_h}\}$, from $t = 4$ to 9 s. Fig. 5.9(a) shows the position (current (solid line) and desired (dashed line)) brought under control while $\boldsymbol{f}_{h_{\text{ext}}}$ is applied thanks to the integral term in (5.20). Similarly in Fig. 5.9(d) a constant external torque disturbance ($\boldsymbol{\tau}_{h_{\text{ext}}} = [0.2\ \ 0.175\ \ 0.15]^T$ Nm) is applied, about the 3 principal axes $\{\vec{\boldsymbol{X}}_{B_h}, \vec{\boldsymbol{Y}}_{B_h}, \vec{\boldsymbol{Z}}_{B_h}\}$, from $t = 12$ to 18 s. Fig. 5.9(b) shows the orientation that gets disturbed by this external torque and brought under control after a short transient, thanks to the integral term in (5.23). The in-zoomed Fig. 5.9(a) shows that the position tracking error is very minimal in powers of 10^{-9}. This simulation provides a first confirmation of the validity of the robustness of the controller during hovering with external disturbance and also the ability of reorienting the hexarotor while maintaining a fixed position, thus showing the 6–DoF capabilities of the hexarotor. This point will also be addressed more thoroughly by the next simulation.

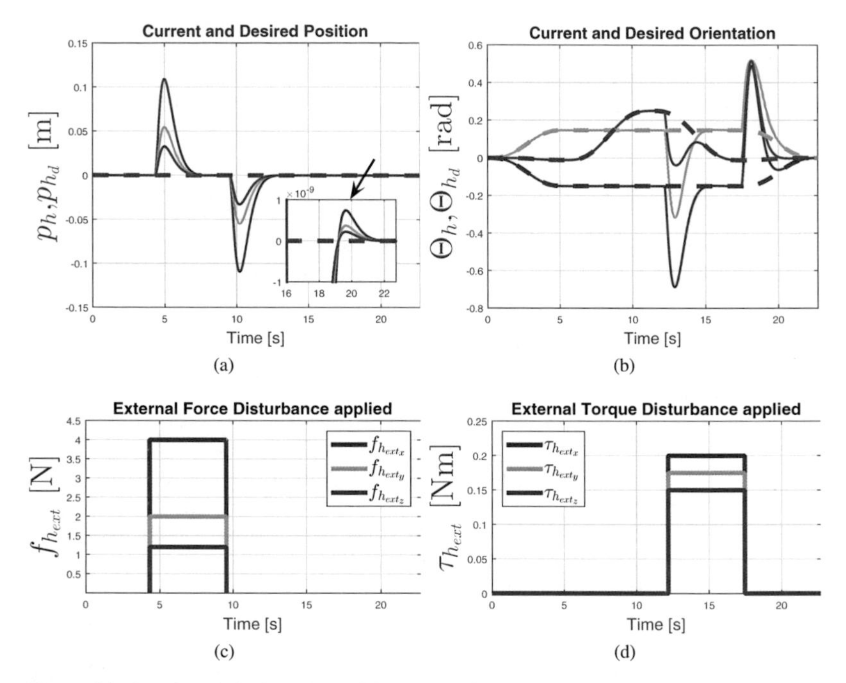

Figure 5.9: Results of the hovering with external force/torque disturbance. 5.9(a): Desired (dashed line) $\boldsymbol{p}_{h_d}$ and current (solid line) position $\boldsymbol{p}_h$ in x(red), y(green) and z(blue). 5.9(b): Desired (dashed line) $\boldsymbol{\Theta}_{h_d}$ and current (solid line) orientation $\boldsymbol{\Theta}_h$ in roll(red), pitch(green) and yaw(blue). 5.9(c–d): external force($\boldsymbol{f}_{h_{\text{ext}}}$) and torque($\boldsymbol{\tau}_{h_{\text{ext}}}$) applied to the hexarotor

5.5.2 6 DoF trajectory tracking

In this simulation, we have addressed a more complex trajectory following a square path with vertexes $\{V_1, V_2, V_3, V_4, V_5, V_6, V_7\}$. Each vertex was associated with the following desired positions and orientations[3]

- $\boldsymbol{V}_1$: $\boldsymbol{p}_{h_d} = [0\,0\,0]^T$, $\boldsymbol{\Theta}_{h_d} = [0^\circ\ \ 0^\circ\ \ 0^\circ]^T$
- $\boldsymbol{V}_2$: $\boldsymbol{p}_{h_d} = [2\,0\,0]^T$, $\boldsymbol{\Theta}_{h_d} = [-18^\circ\ \ 0^\circ\ \ 0^\circ]^T$
- $\boldsymbol{V}_3$: $\boldsymbol{p}_{h_d} = [2\,3\,0]^T$, $\boldsymbol{\Theta}_{h_d} = [-18^\circ\ \ 12^\circ\ \ 0^\circ]^T$
- $\boldsymbol{V}_4$: $\boldsymbol{p}_{h_d} = [2\,3\,1]^T$, $\boldsymbol{\Theta}_{h_d} = [-18^\circ\ \ 12^\circ\ \ 9^\circ]^T$
- $\boldsymbol{V}_5$: $\boldsymbol{p}_{h_d} = [2\,0\,1]^T$, $\boldsymbol{\Theta}_{h_d} = [-18^\circ\ \ 12^\circ\ \ 0^\circ]^T$
- $\boldsymbol{V}_6$: $\boldsymbol{p}_{h_d} = [2\,0\,0]^T$, $\boldsymbol{\Theta}_{h_d} = [-18^\circ\ \ 0^\circ\ \ 0^\circ]^T$
- $\boldsymbol{V}_7$: $\boldsymbol{p}_{h_d} = [0\,0\,0]^T$, $\boldsymbol{\Theta}_{h_d} = [0^\circ\ \ 0^\circ\ \ 0^\circ]^T$

which were traveled along with rest-to-rest motions with maximum linear/angular velocities of 0.3 m/s and 15° /s, and maximum linear/angular accelerations of 0.2 m/s^2 and 5° /s. Figures 5.10(a–d) show the desired trajectory $(\boldsymbol{p}_{h_d}(t), \boldsymbol{\Theta}_{h_d}(t))$, and the tracking errors $(\boldsymbol{e}_{h_p}(t), \boldsymbol{e}_{h_R}(t))$. The same initial condition as in Sec. 5.5.1 is considered. The optimized value of $\alpha_h' = 26.5^\circ$ and $\beta_h' = 19^\circ$ for this trajectory obtained from Sec. 5.4.2 has been used. Here it is clearly illustrated that at the vertex $\boldsymbol{V}_4$ the hexarotor exploits the 6 DoFs. Clearly, the 6-DoF is excluded from the capabilities of the underactuated quadrotor. Note again how the tracking errors are kept to minimum (in power of 10^{-5}) despite the more complex motion involving several reorientations of the propellers. This confirms again the validity of the proposed controller.

5.6 Hexarotor Prototype

In this section, we explain in detail the design and manufacturing of a prototype of the developed fully actuated tilted propeller hexarotor. Some experimental results will be shown to demonstrate the full actuation properties. The assembled prototype is shown in Fig. 5.11(a).

[3] Here, for the sake of clarity, we represent orientations by means of the classical roll/pitch/yaw Euler set $\boldsymbol{\Theta}_h \in \mathbb{R}^3$.

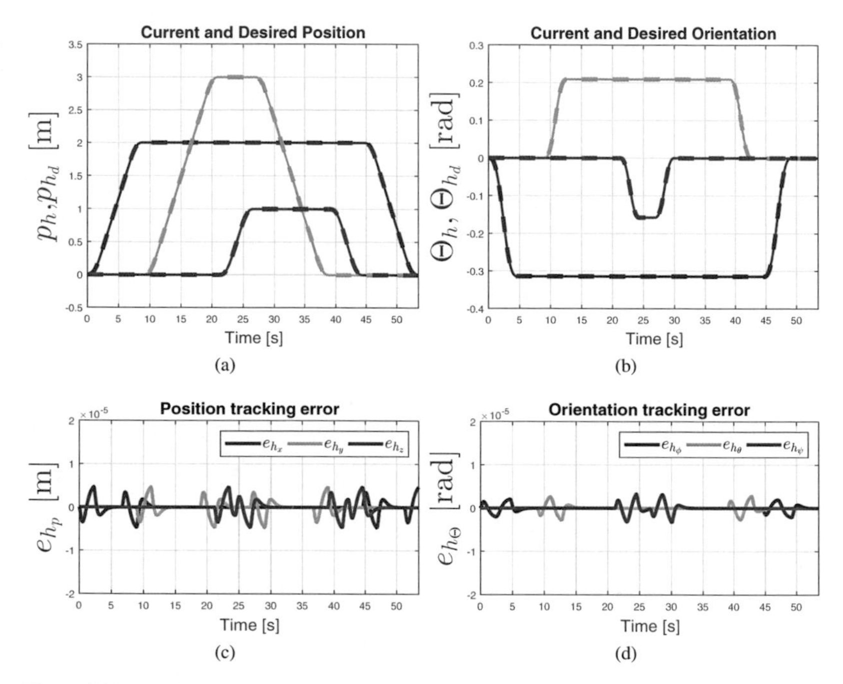

Figure 5.10: Results of the robust 6 DoFs trajectory tracking. 5.10(a): Desired (dashed line) $\boldsymbol{p}_{h_d}$ and current (solid line) position $\boldsymbol{p}_h$ in x(red), y(green) and z(blue). 5.10(b): Desired (dashed line) $\boldsymbol{\Theta}_{h_d}$ and current (solid line) orientation $\boldsymbol{\Theta}_h$ in roll(red), pitch(green) and yaw(blue). 5.10(c–d): behavior of the position/orientation tracking errors $(\boldsymbol{e}_{h_p}, \boldsymbol{e}_{h_\Theta})$.

(a) (b)

(c) (d)

Figure 5.11: Hardware setup of the tilted propeller hexarotor. 5.11(a): Complete assembled hexarotor prototype setup with tilted propeller, tilting adaptor, flight control boards and other hardware. 5.11(b): Zoomed in figure which shows the individual electronic parts in the hexarotor which consists of flight controller, brushless controller (BL-Ctrl), BL-Ctrl access port, vicon markers, power board, landers and arms. 5.11(c): Propeller tilting adaptor which can rotate w.r.t. all the three axes $(\vec{\boldsymbol{X}}_{P_{hi}}, \vec{\boldsymbol{Y}}_{P_{hi}}, \vec{\boldsymbol{Z}}_{P_{hi}})$ while maintaining $\boldsymbol{O}_{P_{hi}}$ at the same location. 5.11(d): Tilting adpator fixed in the arm mounting the brushless motor and propeller setup.

5.6.1 Hardware

The hexarotor hardware consist of 6 aluminum arms, each 35.5 cm long and weighing about 20 g, that are supported by carbon fiber base body circular frame. Separated from the frame below through dampers are fixed the brushless motor power distribution board that mounts six optimized brushless controllers of version BL-Ctrl V2.0 from mikrokopter. The advantage of this version is that, it gives the temperature and voltage measurement which is fed back to the flight controller with a 11-bit resolution with the peak current of 35 A. These measurements are useful to implement a direct motor velocity control (as explained later in Sec. 5.6.2), in order to have a closed loop control independent of battery voltage.

A flight controller FlightCtrl V2.5 from mikrokopter which has air pressure sensor and altitude sensor that work up to 5000 m is fixed on top of the power board. This controller also mounts the IMU composed of two 3-axis analog sensors: an accelerometer with measurement range of $\pm 2\ g$ and a gyroscope with measurement range $\pm 300\ deg/s$, both read with a 10-bit analog to digital converter. The board communicates with the brushless motor controllers through a standard I^2C bus. Furthermore, the controller is connected to a remote PC with through a serial connection. The control inputs, consisting of propeller velocities, are generated on the remote PC. Vicon markers are placed on top of the controller boards so that the hexarotor can be localized using the motor capture system.

Each arm of the hexarotor has an MK3640 brushless motors which has a no-load speed of 500 rpm/V with the maximum load current of 30 A. $13''$ propellers are attached to the motors through 5 mm bore mount. The motor-propeller setup is fixed atop of the tilted propeller adaptor which is shown in Fig. 5.11(d). The adaptors, which are inserted between the motors and the arms, are used to select the proper tilt of each motor. They are custom manufactured has two movable cicular arcs mounted on a rotating disc as shown in Fig. 5.11(c). The circular arcs can move within the C-shaped arc to create a α_h and β_h tilt as required, with a range of $\pm 30\ deg$. Therefore, the arcs provide rotation about $\vec{\boldsymbol{X}}_{P_{hi}}$ and $\vec{\boldsymbol{Y}}_{P_{hi}}$. In addition, the rotating disc on the base rotate 360 deg about $\vec{\boldsymbol{Z}}_{P_{hi}}$, providing altogether 3D rotation capabilities. Figure. 5.11(d) shows the motor-propeller mounted in the tilting adaptor. Note that that the radii of the cicular arcs are constructed so that the center of the propeller frame $\boldsymbol{O}_{P_{hi}}$ always lie at the same point irrespective of the selected tilt angle. This ensures that all the $\boldsymbol{O}_{P_{hi}}$ for $i = 1 \ldots 6$ lie in the same plane therefore maintaining a symmetric configuration. The distance between propeller arms and $\boldsymbol{O}_{P_{hi}}$ is 10.5 cm. Therefore the force from each propellers i is generated from the fixed $\boldsymbol{O}_{P_{hi}}$ and the same plane.

The platform is powered by LiPo battery with capacity of 4500 $mAh/35\ C$ which has 6 cell and provides approximately 15 min of flight time. With all the hardware assembled, the total mass of the hexarotor is 2.95 Kg. The state of the UAV in our $10m \times 10m$ flying arena is provided at 120 Hz by a motion capture system (VICON), which is also used to collect ground truth data.

5.6.2 Software

The main control and estimation algorithms are performed in the base station, which is a ROS enabled Ubuntu 14.04 PC. The Telekyb software framework (Grabe *et al.*, 2013) which is an open source architecture developed here at MPI for mobile robots is utilized for the functioning and coordination of the different components of the tilted propeller hexarotor. Telekyb is based on the Robot Operating System (ROS), a software framework for robot, which provides the functionality similar to an operating system such as control of various hardware devices, transferring messages, commanding services, running processes, etc. The processes, represented in a graph architecture, run in nodes that subscribe or publish data coming from a variety of hardware (sensors, actuators, microcontrollers, etc.,) as well as software (controller, state estimator, etc.,) components. All the developed software components such as the controller design, state estimation, trajectory generator, hardware interfaces, etc., are compatible with both ROS and Telekyb. All the nodes have the ability to communicate between one another.

In the system dynamics and controller design mentioned in Sec. 5.2 and Sec. 5.3 respectively, control inputs $\boldsymbol{u}_h$ are related to the propeller velocity $\bar{\omega}_{hi}^2$ as seen in (5.11). In almost all the UAV setup used until now in research domain, the propeller speeds are controlled by varying the duty cycle of the input voltage from the battery according to the generated control input. This open-loop transfer of control inputs are implemented in the FlightCtrl which is efficient until the battery volts start to drop down. As the input voltage from the power source starts to drop the strength of the modulated signal also falls accordingly thereby reducing the propeller velocity without the knowledge of the controller.

In order to solve this problem, it is better to directly control the brushless motor through a closed loop architecture, where the propeller velocity is maintained irrespective of the input voltage. Therefore in the hexarotor prototype, we have installed the recently developed open source software[4] from LAAS CNRS. Through this control architecture, both the BL-Ctrl and the FltrCtrl software is overwritten to implement a closed loop brushless motor control. The velocity controller implemented here gives a precise control of the spinning velocity of the propellers. Moreover, there are safety features added to limit the current when the maximum peak current is attained. The BL-Ctrl is interfaced with the Telekyb software framework in which the controller resides and therefore passes the exact control input to the motors irrespective of the input voltage.

5.7 Experimental Validation

In this section, we present two experiments on the tilted propeller hexarotor to test the performance characteristics of the fully actuated UAV. As the main objective of previous sections were to model, design and develop the hexarotor prototype, this section is pri-

[4]https://git.openrobots.org/projects/tk3-mikrokopter

marily meant to provide insight on the capability and the feasibility of the generation of 6DoF to position as well orient a UAV in real world. In order to check the features of the proposed UAV, we test it with: (i) hovering and reorienting; and (ii) 6DoF trajectory tracking.

Our aim is to show how the tilted propeller hexarotor behavior differs from the standard quadrotor while performing the tasks. The controller parameters and the design parameters are fixed throughout the experiment. These experiments were not designed to test the optimized angle of tilt but the main capabilities of the hexarotor itself. Therefore, the tilt angles were tuned to avoid any singularity effect: $\boldsymbol{\alpha}_h = 0.436332\ rad$ and $\boldsymbol{\beta}_h = 0.261799\ rad$. Note that the alternate propellers are tilted in opposite directions to maintain the symmetric stability configuration. The inertial parameters were found through the exact CAD model of the prototype. The Table. 5.2 lists all the experimental parameters.

5.7.1 Hovering and Reorienting

In this experiment, the objective is to allow the hexarotor to hover at a certain height of $z_h = 1\ m$ and then reorient the UAV while still maintaining the position. Clearly this task is impossible with a standard quadrotor, in which a non-zero roll or pitch causes a movement in X or Y-axis respectively. As seen in Fig. 5.12(a), the UAV takes off at $t < 5\ s$ and hovers at a position $\boldsymbol{p}_h = [0\ 0\ 1]^T\ m$. Initially during take-off, there is a uncommanded yaw rotation experienced by the hexarotor which the controller forces to recover as can be seen in Fig. 5.12(b)(green). This behavior is due to mismatches in the inertial parameters and minor difference in the tilt angle mechanical installation. This accumulated error in the yaw is recovered by the controller before the reorientation angles are instructed to the UAV. At $t = 20\ s$, the UAV is commanded a roll of $\phi_{h_d} = 12\ deg = 0.2\ rad$ as can be seen in Fig. 5.12(b) (red dashed). Similarly at $t = 25\ s$ and $t = 30\ s$, a pitch of $\theta_{h_d} = -12\ deg = -0.2\ rad$ (blue dashed) and a yaw of $\psi_{h_d} = -30\ deg = -0.52\ rad$ (black dashed) are commanded respectively. The hexarotor follows the desired orientation for the ϕ_h, θ_h and ψ_h as can be seen in red, blue and green respectively in Fig. 5.12(b). It can be observed that the roll and pitch tracking is very slow. This is attributed to the fine tuning that needs to improved to have robust tracking performance. All throughout the reorientation of the hexarotor, the position $\boldsymbol{p}_h$ is maintained at the desired hovering as shown in Fig. 5.12(a). The hexarotor is commanded to do the reorientation twice.

Figure 5.12(c) and Figure 5.12(d) show the linear and angular velocity of the tilted propeller hexarotor prototype respectively. There are no big spikes but noisy measurements from the sensors. The experiment proves the full actuation of the UAV with possibility of the hexarotor to reorient itself while still maintaining the same position. Figure 5.12(e) and Figure 5.12(f) show the position and orientation error respectively during this task. The position error is minimal (less than 5 cm) while the hexarotor was oriented with roll and pitch at 12 deg. Since the orientation tracking is slower as highlighted earlier, the plot shows that the error jumps up to 0.1 rad before coming back closer to zero. This

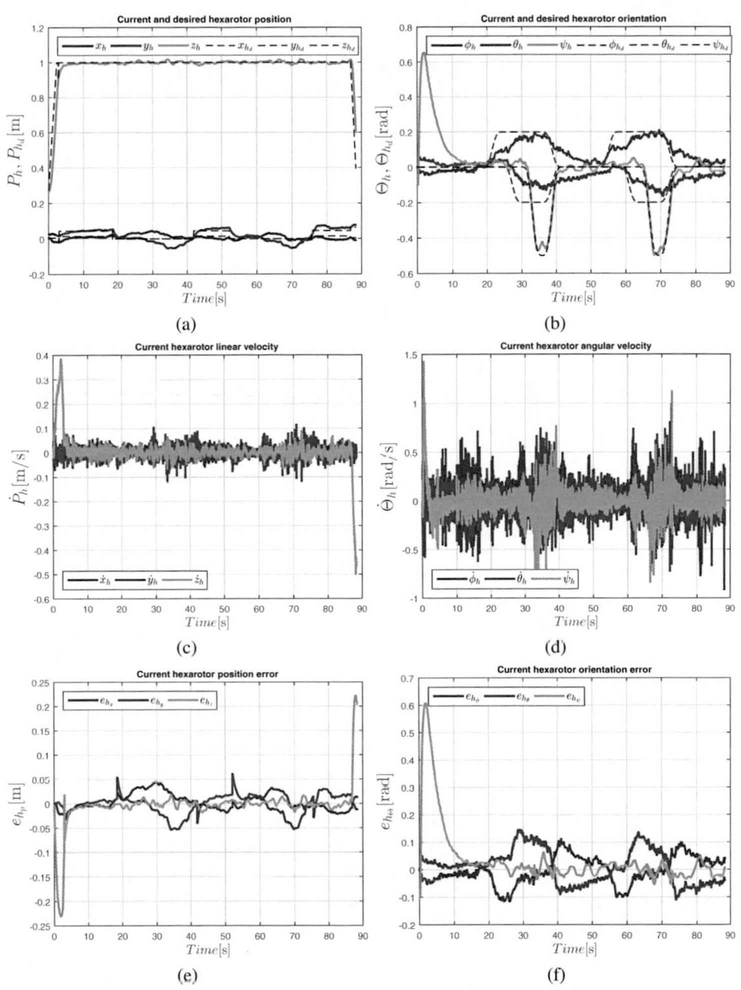

Figure 5.12: Tilted propeller hexarotor - Hovering and reorienting experiment. 5.12(a) actual position $\boldsymbol{p}_h$ in x_h(red), y_h(blue), z_h(green) and desired position $\boldsymbol{p}_{h_d}$ (black dashed) of the UAV; 5.12(b) actual orientation $\boldsymbol{\Theta}_h$ in ϕ_h(red), θ_h(blue), ψ_h(green) and desired orientation $\boldsymbol{\Theta}_{h_d}$ in ϕ_{h_d}(red dashed), θ_{h_d}(blue dashed), ψ_{h_d}(green dashed) of the UAV; 5.12(c): actual linear velocity $\dot{\boldsymbol{p}}_h$ in $\dot{x}_h$(red), $\dot{y}_h$(blue) and $\dot{z}_h$(green) of the UAV; 5.12(d) actual angular velocity $\dot{\boldsymbol{\Theta}}_h$ in $\dot{\phi}_h$(red), $\dot{\theta}_h$(blue), $\dot{\psi}_h$(green) of the UAV; 5.12(e): position tracking error e_{h_p} in e_{h_x}(red), e_{h_y}(blue) and e_{h_z}(green). 5.12(f): orientation tracking error $e_{h\Theta}$ in e_{h_ϕ}(red), $e_{h\theta}$(blue) and e_{h_ψ}(green);

Parameter	Description	Value	Unit
m_h	mass of the hexarotor	2.925	Kg
g	gravity acceleration	9.81	m/s^2
I_{xx_h}	inertia along X-axis	0.099	$Kg.m^2$
I_{yy_h}	inertia along Y-axis	0.098	$Kg.m^2$
I_{zz_h}	inertia along Z-axis	0.191	$Kg.m^2$
k_f	lift coefficient	$1.27 * 10^{-3}$	N/Ω^2
k_m	drag coefficient	$2.45 * 10^{-5}$	Nm/Ω^2
$\boldsymbol{L}_{x_h}$	arm length	0.355	m
$\boldsymbol{\alpha}_h$	propeller tilt w.r.t. X-axis	0.436332	rad
$\boldsymbol{\beta}_h$	propeller tilt w.r.t. Y-axis	0.261799	rad
$\boldsymbol{K}_{h_{p1}}$	derivative position error gain	$diag[8,\ 8,\ 5]$	-
$\boldsymbol{K}_{h_{p2}}$	proportional position error gain	$diag[10,\ 10,\ 10]$	-
$\boldsymbol{K}_{h_{p3}}$	integral position error gain	$diag[1,1,2]$	-
$\boldsymbol{K}_{h_{R1}}$	derivative orientation error gain	$diag[14,16,12]$	-
$\boldsymbol{K}_{h_{R2}}$	proportional orientation error gain	$diag[30,37,20]$	-
$\boldsymbol{K}_{h_{R3}}$	integral orientation error gain	$diag[4,8,5]$	-

Table 5.2: Experimental parameters of fully actuated tilted propeller hexarotor.

could be due to thee the tuning and the mismatch in physical parameters of the hexarotor.

The propeller velocity during this experiment is shown in Fig. 5.14(a). Although in general the velocities stayed within the limits, propeller $p3$ (green) shows some spikes. This could be because of the mechanical wear and tear of the adaptors on which the propellers are mounted. Figure 5.15 shows the tilted propeller hexarotor in reoriented position during this hovering experiment.

5.7.2 6 DoF Trajectory Tracking

In the second experiment, we drive the hexarotor to track a 6DoF trajectory which consists of the 3D desired position and 3D desired orientation. The UAV takes off at $t < 5\ s$ and initially hovers at a position $\boldsymbol{p}_h = [0\ 0\ 1]^T\ m$ (Fig. 5.13(a)). During the time of take-off, a yaw rotation is experienced by the hexarotor which it immediately recovers as seen in Fig. 5.13(b). As mentioned earlier, this is due to the minor differences in the tilt angle during the hardware setup. At time $t > 15\ s$, a desired roll of $\phi_{h_d} = 0.2\ rad$ is commanded. Similarly at $t > 25\ s$ and $t > 30\ s$, pitch and yaw are commanded respectively as can be seen in Fig. 5.13(b). This orientation is tracked as desired, thanks to the exact feedback linearization controller. At $t > 20$, x_{h_d} and y_{h_d} of $1\ m$ and $-1\ m$ respectively is given by the trajectory generator. As seen in Fig. 5.13(a), a robust position tracking is carried out by the hexarotor. The full actuation effect could be visibly seen at $30 \leq t \geq 40\ s$ in Fig. 5.13(a) and Fig. 5.13(b), when the hexarotor follows a 6DoF trajectory.

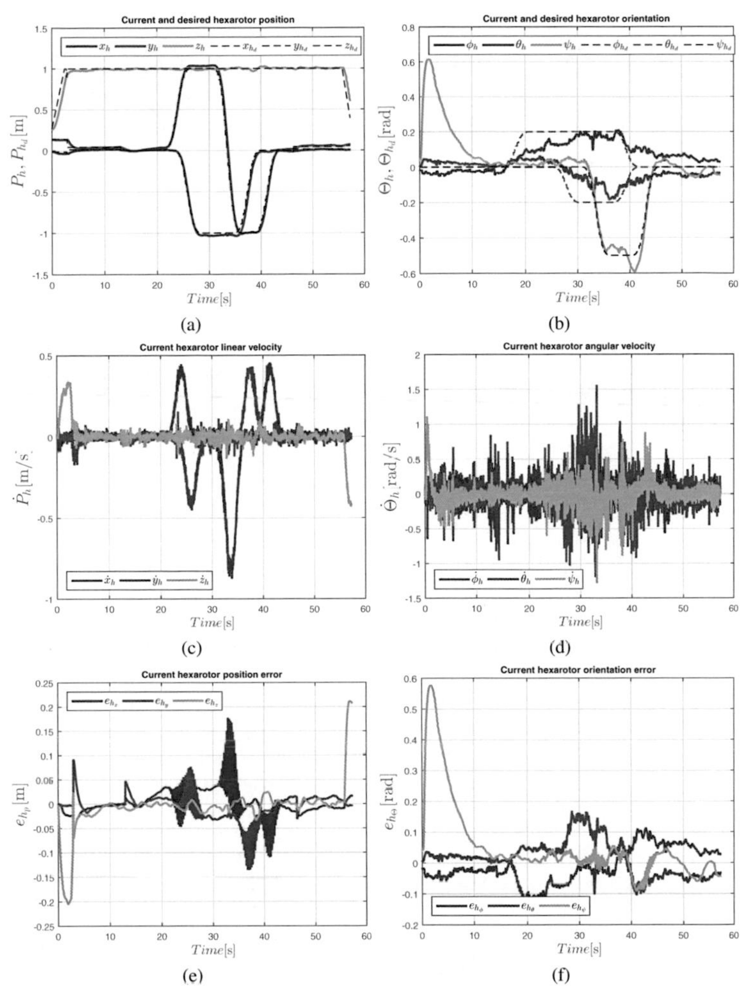

Figure 5.13: Tilted propeller hexarotor - 6DoF Trajectory Tracking Experiment. 5.13(a) actual position $\boldsymbol{p}_h$ in x_h(red), y_h(blue), z_h(green) and desired position $\boldsymbol{p}_{h_d}$ (black dashed) of the UAV; 5.13(b) actual orientation $\boldsymbol{\Theta}_h$ in ϕ_h(red), θ_h(blue), ψ_h(green) and desired orientation $\boldsymbol{\Theta}_{h_d}$ in ϕ_{h_d}(red dashed), θ_{h_d}(blue dashed), ψ_{h_d}(green dashed) of the UAV; 5.13(c): actual linear velocity $\dot{\boldsymbol{p}}_h$ in $\dot{x}_h$(red), $\dot{y}_h$(blue) and $\dot{z}_h$(green) of the UAV; 5.13(d) actual angular velocity $\dot{\boldsymbol{\Theta}}_h$ in $\dot{\phi}_h$(red), $\dot{\theta}_h$(blue), $\dot{\psi}_h$(green) of the UAV; 5.13(e): position tracking error e_{h_p} in e_{h_x}(red), e_{h_y}(blue) and e_{h_z}(green). 5.13(f): orientation tracking error e_{h_Θ} in e_{h_ϕ}(red), e_{h_θ}(blue) and e_{h_ψ}(green);

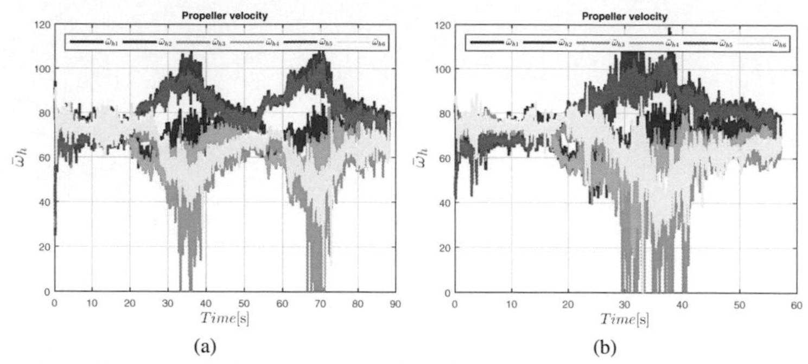

(a) (b)

Figure 5.14: Tilted propeller hexarotor - Propeller velocity during hovering and 6DoF trajectory tracking experiment. 5.14(a) tilted propeller velocity during hovering and reorienting experiment $\bar{\omega}_h$ in $\bar{\omega}_{h1}$(red), $\bar{\omega}_{h2}$(blue), $\bar{\omega}_{h3}$(green), $\bar{\omega}_{h4}$(cyan), $\bar{\omega}_{h5}$(magenta) and $\bar{\omega}_{h6}$(yellow) ; 5.14(b) tilted propeller velocity during trajectory tracking experiment $\bar{\omega}_h$ in $\bar{\omega}_{h1}$(red), $\bar{\omega}_{h2}$(blue), $\bar{\omega}_{h3}$(green), $\bar{\omega}_{h4}$(cyan), $\bar{\omega}_{h5}$(magenta) and $\bar{\omega}_{h6}$(yellow) ;

Figure 5.13(c) and Figure 5.13(d) show the linear and angular velocities of the hexarotor respectively. The linear velocity reaches about 1 m/s, which is a reasonable value. Similarly, the maximum angular velocity is more than 1 rad/s. Figure 5.13(e) and Figure 5.13(f) show the position and orientation error respectively. The position error in general is less than 3 cm which shows the robust trajectory tracking. At $t = 31\ s$, it can be seen that the error started at around 15 cm before coming to zero. This is because of the waypoint which was commanded to a far away distance (2 m) as can be seen in Fig. 5.13(a) during this time. The orientation error has started at 0.1 rad and fallen back to zero. The initial high error is due to the slow response of the controller action.

The propeller velocity during the trajectory tracking experiment is shown in Fig. 5.14(b). Similarly to the other experiment, although the velocities stayed within the limits, propeller $p3$ (green) shows some spikes, probably due to the mechanical wear and tear of the adaptors on which the propellers are mounted.

5.8 Discussions and Possible Extensions

The objective of this chapter is to solve the underactuation problem that affects in standard quadrotors. The limitations of previous 6DoF UAV designs (Sec. 5.1.1) were taken into account while addressing this issue. Summarizing the important development in this chapter:

1. it was proposed a tilted propeller hexarotor to overcome the problem of underactuation is standard UAVs, where the propellers can be rotated both w.r.t. X-axis and Y-axis;

Figure 5.15: Snapshot of tilted propeller hexarotor during hovering and reorientation experiment.

2. it was derived the translational and rotational dynamics of the proposed hexarotor in Sec. 5.2;

3. it was designed an exact feedback linearization and decoupling control for the tilted propeller hexarotor utilizing the availability of 6 control inputs in Sec. 5.3;

4. an optimization study (Sec. 5.4.2) was done to identify the tilt angle that would reduce the power consumption in general so that the flight time of the hexarotor can be increased;

5. it was developed a prototype of the the tilted propeller hexarotor (Sec. 5.6) with an adaptor design to tilt the propellers. Successful experiments have been conducted of UAV reorienting while hovering and 6DoF trajectory tracking, as detailed in Sec. 5.7.

As seen in the experimental validation Sec. 5.7, slow response has been observed for the roll and pitch orientation. We attribute this to tuning and mismatches in inertial parameters, which could be further improved. Further study could also be done to see the performance characteristics for asymmetric propeller tilt angles. From the application point of view, the hexarotor could be used as an UAV platform for aerial manipulation through the installation of light weight manipulators. Furthermore, as mentioned in Chapter. 3, this platform could be used for the human-UAV physical interaction where the torque could also be exchanged along with the force. This will be our research topic in Chapter. 6.

Chapter 6

Human-UAV Physical Interaction with a Fully Actuated UAV

From the discussions until now, it is clear that the external wrench estimation works efficiently and can be employed for a real time UAV application without the inclusion of any additional payload or sensors (Chapter. 2). This have paved the way for the interaction wrench estimation as well leading to the development of novel hardware architecture (Chapter. 3) and software framework for the idea of human-UAV physical interaction. The theoretical development and experiments with UAVs led to identify the major roadblocks that are present for human-UAV physical interaction with the current state-of-art. The need for a robust disturbance rejection controller when the application scenario is moving towards unknown outdoor environment made us to propose a non-linear controller in Chapter. 4. The other major drawback was the underactuation of quadrotors, which restricted the wrench interaction between humans and UAVs, limiting the interaction to only intuitive exchange of forces but not torques. In order to address this issue, a novel fully actuated tilted propeller hexarotor has been proposed in Chapter. 5.

All the above developments of new methodologies, implementation techniques, software framework, hardware architecture, controller designs and novel UAV models, opens new roads of possibilities and applications. Some new questions which arise then are: *Is it possible to implement the external interaction wrench estimator for a human-UAV physical interaction application utilizing a fully-actuated UAV? Can the proposed robust adaptive super twisting sliding mode controller be implemented for such a system? Is there a feasible hardware architecture for this setup? How would the system dynamics behave? Can a human exchange torque with UAVs in this scenario without compromising the stability?*

In this chapter, we answer these questions by initially developing a dynamic UAV model for a generic $n \geq 6$ non-coplanar propellers to obtain a fully actuated UAV and then implement the external interaction wrench estimator discussed earlier in Chapter. 2 in this setup. Later, the admittance control framework for both force and torque are defined for human-UAV physical interaction. Thanks to the fully actuated dynamics, here it is possible to exchange interaction torques along with the forces between the humans and UAVs. The robust adaptive super twisting controller is adapted to the dynamical

Figure 6.1: Standard quadrotor setup for UAV-HRPI.

model of the fully actuated UAV.

Note that the latest improvements and the discussion presented in this chapter is submitted in Rajappa *et al.* (2017b).

6.1 Introduction

As mentioned clearly in Chapter. 3, Human-Robot Interaction (HRI) with aerial robots is a recent research domain. The use of UAV for aerial interaction and manipulation task has started to pickup in the last couple of years. The application domain is quiet varied from outdoor non-human accessible dangerous situation (e.g., stress break repair in wind turbine blade, nuclear facility maintenance, etc.) to indoor scenario (e.g., peg-in-hole, UAV painting, etc.). However, with the current trend of hardware miniaturization and simultaneous increase in computational power, advancements in computer vision and control techniques, UAV applications in human-populated areas will start to be considered feasible. In this context, new interaction paradigms between humans and UAVs should be developed, as proposed earlier in chapter. 3.

First, detecting and properly reacting to mutually applied forces will be fundamental to increase safety when humans and UAVs share the same space. In addition, many foreseeable applications will involve some form of intentional force exchange due to physical contact (i.e., physical interaction). UAV systems for goods transportation or tool deployment to workers may involve the recipient to pick-up the transported object

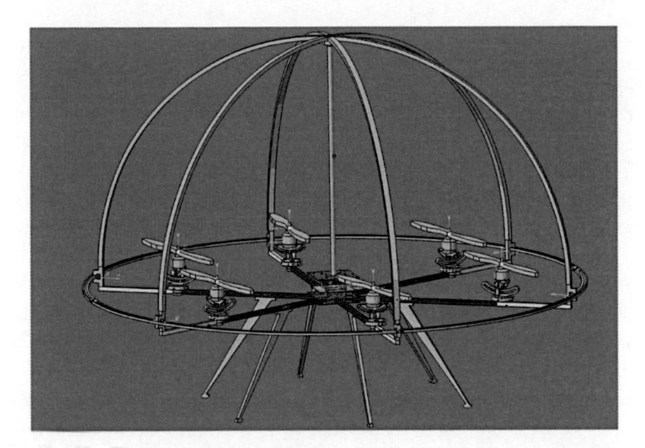

Figure 6.2: CAD model of UAV-HRPI setup with a fully actuated tilted propeller hexarotor with the tilting adaptor, sensor ring, interaction surface, safety arm and other standard UAV hardware.

directly from the UAV when there is no available landing or deployment area. Moreover, kinesthetic trajectory teaching will allow untrained users setting up general purpose UAV systems.

In Chapter. 3, we addressed the main reasons for the lack of research in UAV-HRPI through the development of hardware setup, software framework along with an interaction surface. During the experiments, we also identified the other concerns that could come into play in a real-time application. When it comes to physical interaction, the underactuated nature of standard quadrotors may become a serious issue that hinders the capabilities and overall stability of the aerial vehicle.

In particular, since the roll and pitch angles of the quadrotor cannot be selected independently w.r.t. the other degrees of freedom, a torque exchange around the longitudinal and latitudinal axes of the UAV cannot result in an appropriate reaction of the quadrotor. Hence, underactuation limits the possible physical interaction to an exchange of forces along the three axes and torques around the vertical axis. Moreover, and more importantly from a safety point of view, if torques around the latitudinal and longitudinal axes are accidentally exchanged, the UAV could become unstable because of the underactuated system dynamics and therefore no more safe for human interaction. In addition, when a force exchange happens, undesired roll and pitch angles may be needed to perform the required action. For example, the quadrotor needs to perform a nonzero pitch to comply with a force applied along its longitudinal axis.

6.1.1 Related Works

As mentioned earlier in Chapter. 3, works have been done previously in the context of Human-Robot Interaction (HRI) with aerial robots. HRI has been explored mostly by considering either intermediary physical interfaces (monitors, joysticks, haptic devices, etc.) or visual (hand gestures, upper body gesture, face tracking, etc.) and auditory sensory channels. Refer Sec. 3.1.1 for the detailed literature review on the new emerging technologies on controller design, mechanical hardware, sensors, UAV models, tilting mechanisms, UAV interaction and interfaces. Several UAV models have been proposed in order to overcome the underactuation issue (Refer Sec. 5.1.1). One such model has been proposed in Chapter. 5 through the mechanism of tilted propellers in a hexarotor. This setup has the advantage of simple control design and stability compared to the tilting mechanisms.

With regards to the controller, several control designs are available in literature (Refer Sec. 4.1.1). The need for a robust controller during a UAV application in an outdoor environment is well known. Here, the main intention is therefore to perform the human-UAV physical interaction with a fully actuated UAV with tilted propeller mechanism utilizing a robust controller proposed in Chapter. 4.

6.1.2 Methodologies

Therefore in this chapter,

1. it is introduced the dynamic model for a generic non-coplanar $n \geq 6$ propeller fully actuated UAV.
2. it is developed a UAV-HRPI scheme (control architecture, interaction estimation algorithm) employing a fully actuated UAV.
3. it is implemented an admittance control framework where the desired trajectory is changed based on the interaction forces and torques.
4. it is derived the adaptive super twisting controller (proved effective against parameter uncertainties and disturbances), for robust trajectory tracking of a fully actuated UAV with fixed tilted non-coplanar propellers.
5. it is compared the UAV-HRPI in case of standard (underactuated) quadrotor and fully actuated UAV.

6.2 Design and Modeling

In Chapter. 5, a fully actuated tilted propeller hexarotor is proposed. Here, it is presented the dynamic model of a generic fully actuated multirotor UAV with fixed non-coplanar

$n \geq 6$ (in order to obtain a fully actuated system) propellers. Though philosophically this should work similar to the hexarotor, the difference is that as the number of propellers are increased, the degrees of freedom are more than 6 (i.e., DoFs ≥ 6). Therefore the 6 DoF trajectroy tracking for a given pose can be tracked with different propeller velocities. Note that the study on the propeller redundancy is not the objective in this case.

The derivation of generic fully actuated multirotor UAV model follows the same conceptual steps followed in Chapter. 5 for a hexarotor with tilted propellers, of which this model is a trivial generalization. Therefore, for the sake of brevity, the full derivation is omitted here and referred to Chapter. 5 or Rajappa *et al.* (2015) for details. Let be $\mathcal{F}_W : \{\boldsymbol{O}_W, \vec{\boldsymbol{X}}_W, \vec{\boldsymbol{Y}}_W, \vec{\boldsymbol{Z}}_W\}$ the world inertial frame, and let be $\mathcal{F}_{B_h} : \{\boldsymbol{O}_{B_h}, \vec{\boldsymbol{X}}_{B_h}, \vec{\boldsymbol{Y}}_{B_h}, \vec{\boldsymbol{Z}}_{B_h}\}$ the body frame attached to the multirotor UAV, where $\boldsymbol{O}_{B_h}$ coincides with its center of mass (CoM). Let the frame associated with the i-th propeller be defined as $\mathcal{F}_{P_{hi}}$: $\{\boldsymbol{O}_{P_{hi}}, \vec{\boldsymbol{X}}_{P_{hi}}, \vec{\boldsymbol{Y}}_{P_{hi}}, \vec{\boldsymbol{Z}}_{P_{hi}}\}$, where $i = 1 \ldots n$ where $n \geq 6$. The frame reference in this generic case is similar as defined in Fig. 5.1. Let $\boldsymbol{p}_h = \begin{bmatrix} x_h & y_h & z_h \end{bmatrix}^T \in \mathbb{R}^3$ describe the position of $\boldsymbol{O}_{B_h}$ in $\mathcal{F}_W$ and let $\boldsymbol{\Theta}_h = \begin{bmatrix} \phi_h & \theta_h & \psi_h \end{bmatrix}^T \in \mathbb{R}^3$ be the standard roll, pitch and yaw angles respectively which describe the orientation of $\mathcal{F}_{B_h}$ in $\mathcal{F}_W$, with $\phi, \theta \in [-\pi/2, \pi/2]$ and $\psi \in [0, 2\pi]$. The basic multirotor states are therefore

$$\begin{bmatrix} \boldsymbol{p}_h^T & \boldsymbol{\Theta}_h^T \end{bmatrix}^T = \begin{bmatrix} x_h & y_h & z_h & \phi_h & \theta_h & \psi_h \end{bmatrix}^T. \tag{6.1}$$

Neglecting the external forces and torques, the translational dynamics of the tilted propeller multirotor UAV based on Newton-Euler formulation can be written similar to hexarotor in (5.12) as

$$\ddot{\boldsymbol{p}}_h = \begin{bmatrix} 0 & 0 & -g \end{bmatrix}^T + \frac{1}{m_h} {}^W\boldsymbol{R}_{B_h} \boldsymbol{F}(\boldsymbol{\alpha}_h, \boldsymbol{\beta}_h, \boldsymbol{\lambda}_h) \boldsymbol{u}_h \tag{6.2}$$

where $\boldsymbol{p}_h \in \mathbb{R}^3$ is the position of $\boldsymbol{O}_{B_h}$ in $\mathcal{F}_W$, g is the acceleration due to gravity, m_h is the mass of the vehicle, ${}^W\boldsymbol{R}_{B_h} \in SO(3)$ is the rotation matrix representing the orientation of $\mathcal{F}_{B_h}$ w.r.t. $\mathcal{F}_W$, $\boldsymbol{u}_h$ is the control input given by

$$\boldsymbol{u}_h = [\bar{\omega}_{h1}^2 \ \bar{\omega}_{h2}^2 \ \bar{\omega}_{h3}^2 \ \ldots \ \bar{\omega}_{hn}^2]^T \in \mathbb{R}^{n \times 1}, \tag{6.3}$$

i.e., the squares of the rotational speeds $\bar{\omega}_{hi} \, \forall i = 1 \ldots n$ of each propeller, $\boldsymbol{F}(\boldsymbol{\alpha}_h, \boldsymbol{\beta}_h, \boldsymbol{\lambda}_h) \in \mathbb{R}^{3 \times n}$ is the matrix that relates $\boldsymbol{u}_h$ with the total thrust produced by the propellers (expressed in body frame), $\boldsymbol{\alpha}_h = (\alpha_{h1}, \alpha_{h2}, \ldots, \alpha_{hi}) \forall i = 1 \rightarrow n$ represents the tilt angle of i-th propeller w.r.t. $\vec{\boldsymbol{X}}_{P_{hi}}$, $\boldsymbol{\beta}_h = (\beta_{h1}, \beta_{h2}, \ldots, \beta_{hi}) \forall i = 1 \rightarrow n$ represents the tilt angle of i-th propeller w.r.t. $\vec{\boldsymbol{Y}}_{P_{hi}}$ and $\boldsymbol{\lambda}_h = (\lambda_{h1}, \lambda_{h2}, \ldots, \lambda_{hi}) \forall i = 1 \rightarrow n$ is the angular direction of the segment $\overline{\boldsymbol{O}_{B_h} \boldsymbol{O}_{P_{hi}}}$ on the $\vec{\boldsymbol{X}}_{B_h} \vec{\boldsymbol{Y}}_{B_h}$ plane.

The rotation dynamics of a multirotor can also be written similar to a hexarotor (5.15) as

$$\dot{\boldsymbol{\omega}}_{B_h} = -\boldsymbol{I}_{B_h}^{-1}(\boldsymbol{\omega}_{B_h} \times \boldsymbol{I}_{B_h}\boldsymbol{\omega}_{B_h}) + \boldsymbol{I}_{B_h}^{-1}\boldsymbol{H}(\boldsymbol{\alpha}_h, \boldsymbol{\beta}_h, \boldsymbol{\lambda}_h, \boldsymbol{L}_{x_h})\boldsymbol{u}_h \tag{6.4}$$

$${}^{W}\dot{\boldsymbol{R}}_{B_h} = {}^{W}\boldsymbol{R}_{B_h}[\boldsymbol{\omega}_{B_h}]_{\wedge} \tag{6.5}$$

with $[\cdot]_{\wedge}$ being the hat operator from $\mathbb{R}^3$ to $so(3)$. Here $\boldsymbol{\omega}_{B_h} \in \mathbb{R}^3$ is the angular velocity of $\mathcal{F}_{B_h}$ w.r.t. $\mathcal{F}_W$ expressed in $\mathcal{F}_{B_h}$, $\boldsymbol{I}_{B_h}$ is the multirotor body inertia matrix, $\boldsymbol{H}(\boldsymbol{\alpha}_h, \boldsymbol{\beta}_h, \boldsymbol{\lambda}_h, \boldsymbol{L}_{x_h}) \in \mathbb{R}^{3\times n}$ is the matrix that relates the input torque $\boldsymbol{\tau}_h$ to the control input $\boldsymbol{u}_h$ and $\boldsymbol{L}_{x_h} = (L_{x_{h1}}, L_{x_{h2}}, \ldots, L_{x_{hi}}) > 0 \ \forall i = 1 \to n$ is the distance between $\boldsymbol{O}_{P_{hi}}$ and $\boldsymbol{O}_{B_h}$.

The system dynamics defined in (6.2) and (6.4) for a multirotor UAV can be further simplified and written in matricial form as

$$\begin{bmatrix} \ddot{\boldsymbol{p}}_h \\ \dot{\boldsymbol{\omega}}_{B_h} \end{bmatrix} = \boldsymbol{f}_h + \boldsymbol{J}(\boldsymbol{\alpha}_h, \boldsymbol{\beta}_h, \boldsymbol{\lambda}_h, \boldsymbol{L}_{x_h})\boldsymbol{u}_h \tag{6.6}$$

where $\boldsymbol{f}_h \in \mathbb{R}^6$ is the drift vector due to the gravity and the matrix $\boldsymbol{J}(\boldsymbol{\alpha}_h, \boldsymbol{\beta}_h, \boldsymbol{\lambda}_h, \boldsymbol{L}_{x_h}) \in \mathbb{R}^{6\times n}$ that decouples the control inputs is given by

$$\boldsymbol{J}(\boldsymbol{\alpha}_h, \boldsymbol{\beta}_h, \boldsymbol{\lambda}_h, \boldsymbol{L}_{x_h}) = \begin{bmatrix} \frac{1}{m_h}{}^{W}\boldsymbol{R}_{B_h} & 0 \\ 0 & \boldsymbol{I}_{B_h}^{-1} \end{bmatrix} \begin{bmatrix} \boldsymbol{F}(\boldsymbol{\alpha}_h, \boldsymbol{\beta}_h, \boldsymbol{\lambda}_h) \\ \boldsymbol{H}(\boldsymbol{\alpha}_h, \boldsymbol{\beta}_h, \boldsymbol{\lambda}_h, \boldsymbol{L}_{x_h}) \end{bmatrix}. \tag{6.7}$$

The Fig. 6.2 show the CAD model of such a fully actuated UAV which is designed along with the interaction surface.

6.3 UAV-HRPI System

Following the ideas proposed earlier for a standard quadrotor in Chapter. 3, here it is implemented the system architecture depicted in Fig. 6.3 for a fully actuated UAV. First, while flying the UAV should be able to estimate the external wrench applied on it. This is carried out by the interaction wrench estimator based on the residual computation, which we have implemented earlier in Sec. 2.3. Assuming that the UAV is moving in the environment following a desired trajectory planned by a trajectory planner, this trajectory is modified by an admittance controller designed to admit the interaction wrench. The resulting trajectory is used as reference for the low level Adaptive Super Twisting controller.

Therefore, in this work we are solving three main technological problem:
Estimation Problem: *Given the system states ($\boldsymbol{p}_h$, $\boldsymbol{\Theta}_h$), its derivatives ($\dot{\boldsymbol{p}}_h$, $\dot{\boldsymbol{\Theta}}_h$), model parameters ($g$, m_h, $\boldsymbol{I}_{B_h}$, $\boldsymbol{\alpha}_h$, $\boldsymbol{\beta}_h$, $\boldsymbol{\lambda}_h$, $\boldsymbol{L}_{x_h}$) and control input ($\boldsymbol{u}_h$), how to estimate the*

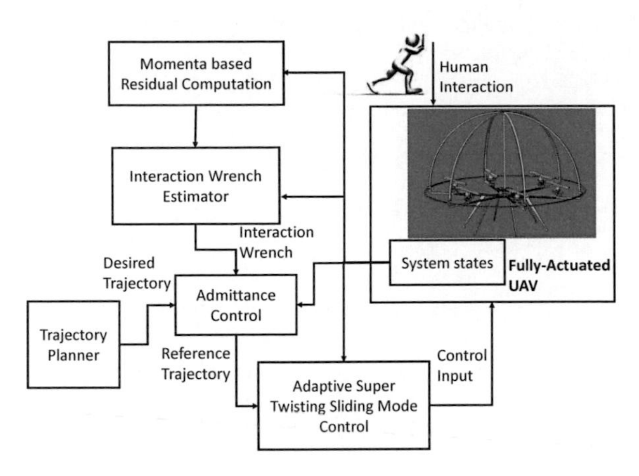

Figure 6.3: System architecture.

human interaction wrenches (forces, torques) acting on the multirotor.
Control Problem I: *Given the system states* ($\boldsymbol{p}_h$, $\boldsymbol{\Theta}_h$), *its derivatives* ($\dot{\boldsymbol{p}}_h$, $\dot{\boldsymbol{\Theta}}_h$), *desired trajectory* ($\boldsymbol{p}_{h_d}(t)$, $\boldsymbol{\Theta}_{h_d}(t)$, $\dot{\boldsymbol{p}}_{h_d}(t)$, $\dot{\boldsymbol{\Theta}}_{h_d}(t)$, $\ddot{\boldsymbol{p}}_{h_d}(t)$, $\ddot{\boldsymbol{\Theta}}_{h_d}(t)$), *model parameters* ($g$, m_h, $\boldsymbol{I}_{B_h}$, $\boldsymbol{\alpha}_h$, $\boldsymbol{\beta}_h$, $\boldsymbol{\lambda}_h$, $\boldsymbol{L}_{x_h}$) *and estimated value of human interaction forces and torques, how to generate the control input so that the human-UAV physical interaction becomes feasible.*
Control Problem II: *Given the system states* ($\boldsymbol{p}_h$, $\boldsymbol{\Theta}_h$), *its derivatives* ($\dot{\boldsymbol{p}}_h$, $\dot{\boldsymbol{\Theta}}_h$), *desired trajectory* ($\boldsymbol{p}_{h_d}(t)$, $\boldsymbol{\Theta}_{h_d}(t)$, $\dot{\boldsymbol{p}}_{h_d}(t)$, $\dot{\boldsymbol{\Theta}}_{h_d}(t)$, $\ddot{\boldsymbol{p}}_{h_d}(t)$, $\ddot{\boldsymbol{\Theta}}_{h_d}(t)$) *and model parameters* (g, m_h, $\boldsymbol{I}_{B_h}$, $\boldsymbol{\alpha}_h$, $\boldsymbol{\beta}_h$, $\boldsymbol{\lambda}_h$, $\boldsymbol{L}_{x_h}$), *how do we solve the output tracking problem of the fully actuated multirotor to track any arbitrary position and orientation using a controller that is effective against both model uncertainties and external perturbations.*

6.3.1 Interaction Wrench Observer

Though derived earlier for a quadrotor UAV earlier in Sec. 2.3, here for the sake of completeness we briefly summarize here its working principle for fully actuated multirotor case. Considering the system states $\zeta_h \triangleq [\dot{x}_h \ \dot{y}_h \ \dot{z}_h \ p \ q \ r]^T$, and following the Lagrangian formulation, the dynamic model of a multirotor in (6.6) can be expressed as

$$\boldsymbol{M}_h\dot{\zeta}_h + \boldsymbol{C}_h(\zeta_h)\zeta_h + \boldsymbol{G}_h = \boldsymbol{\Lambda}_h + \boldsymbol{\Lambda}_{h_{int}} \tag{6.8}$$

where $\boldsymbol{M}_h = \begin{pmatrix} m_h\boldsymbol{I}_3 & \boldsymbol{0}_3 \\ \boldsymbol{0}_3 & \boldsymbol{I}_{B_h} \end{pmatrix} \in \mathbb{R}^{6\times 6}$ is the positive definite inertial matrix, the gravitational vector is $\boldsymbol{G}_h = [0 \ 0 \ m_h g \ 0 \ 0 \ 0]^T$, $\boldsymbol{C}_h(\zeta_h) = \begin{pmatrix} \boldsymbol{0}_3 & \boldsymbol{0}_3 \\ \boldsymbol{0}_3 & -S(\boldsymbol{I}_{B_h}\boldsymbol{\omega}_{B_h}) \end{pmatrix} \in \mathbb{R}^{6\times 6}$ expresses the coriolis and centrifugal terms, $\boldsymbol{\Lambda}_h = \boldsymbol{J}(\boldsymbol{\alpha}_h, \boldsymbol{\beta}_h, \boldsymbol{\lambda}_h, \boldsymbol{L}_{x_h})\boldsymbol{u}_h$ obtained from (6.6) is the nominal wrench due to the control input and $\boldsymbol{\Lambda}_{h_{int}} = [\boldsymbol{F}_{h_{int}}^T \quad \boldsymbol{\tau}_{h_{int}}^T]^T \in \mathbb{R}^6$ is the external human interaction wrench acting on the multirotor.

The residual-based observer relies on the idea of the generalized momenta $\boldsymbol{Q}_h = \boldsymbol{M}_h \zeta_h$. The first-order dynamic equation for the momentum is

$$\dot{\boldsymbol{Q}}_h = \boldsymbol{\Lambda}_h + \boldsymbol{\Lambda}_{h_{int}} + \boldsymbol{C}_h^T(\zeta_h)\zeta_h - \boldsymbol{G}_h. \tag{6.9}$$

We define the residual vector $\boldsymbol{r}_h \in \mathbb{R}^6$ for the interaction wrench estimation of the multirotor as

$$\boldsymbol{r}_h(t) = \boldsymbol{K}_h\left(\boldsymbol{Q}_h - \int_0^t (\boldsymbol{\Lambda}_h + \boldsymbol{C}_h^T(\zeta_h)\zeta_h - \boldsymbol{G}_h + \boldsymbol{r}_h)ds\right), \tag{6.10}$$

where $\boldsymbol{K}_h > 0$ is the diagonal gain matrix. The dynamic evolution of residual $\boldsymbol{r}_h$ satisfies

$$\dot{\boldsymbol{r}}_h = \boldsymbol{K}_h(\boldsymbol{\Lambda}_{h_{int}} - \boldsymbol{r}_h), \quad \text{when } \boldsymbol{r}_h(0) = 0, \tag{6.11}$$

which has an exponentially stable equilibrium at $\boldsymbol{r}_h = \boldsymbol{\Lambda}_{h_{int}}$. For "sufficiently" large gains the dynamic residual in (6.11) becomes

$$\boldsymbol{r}_h \simeq \boldsymbol{\Lambda}_{h_{int}}. \tag{6.12}$$

Being a model-based estimating approach, if a particular component of $\boldsymbol{\Lambda}_{h_{int}}$ is zero then the scalar $\boldsymbol{r}_h$ corresponding to that component converges to zero. Hence, the estimated external wrench is

$$\widehat{\boldsymbol{\Lambda}}_{h_{int}} = \begin{bmatrix} \widehat{\boldsymbol{F}}_{h_{int}} \\ \widehat{\boldsymbol{\tau}}_{h_{int}} \end{bmatrix} = \boldsymbol{r}_h, \tag{6.13}$$

where $\widehat{*}$ indicates the estimated value of a variable $*$. Refer Sec. 2.3 for the detailed explanation of the observer properties.

6.3.2 Admittance Control

In the admittance control framework the desired trajectory $\boldsymbol{p}_{h_d}(t)$, $\dot{\boldsymbol{p}}_{h_d}(t)$, $\ddot{\boldsymbol{p}}_{h_d}(t)$ in $\mathcal{F}_W$ is modified based on the estimated interaction wrenches $\widehat{\boldsymbol{\Lambda}}_{h_{int}}$ to provide a reference trajectory $\boldsymbol{p}_{h_a}(t)$, $\dot{\boldsymbol{p}}_{h_a}(t)$, $\ddot{\boldsymbol{p}}_{h_a}(t)$ to the low level controller. Let the admittance wrench $\boldsymbol{\Lambda}_{h_a}$ be written as

$$\boldsymbol{\Lambda}_{h_a} = \begin{bmatrix} \boldsymbol{F}_{h_a} \\ \boldsymbol{\tau}_{h_a} \end{bmatrix} = \begin{bmatrix} {}^W\boldsymbol{R}_{B_h} & \boldsymbol{0}_{3\times3} \\ \boldsymbol{0}_{3\times3} & \boldsymbol{0}_{3\times3} \end{bmatrix} \begin{bmatrix} \widehat{\boldsymbol{F}}_{h_{int}} \\ \widehat{\boldsymbol{\tau}}_{h_{int}} \end{bmatrix}, \tag{6.14}$$

with the admittance force expressed in $\mathcal{F}_W$ and torque in $\mathcal{F}_{B_h}$.

In order to modify the desired trajectory, we consider the UAV as an ideal mass-spring-damper system driven by the state equation for position and orientation

$$\ddot{\boldsymbol{p}}_{h_a}=\frac{\boldsymbol{F}_{h_a}+D_F(\dot{\boldsymbol{p}}_{h_d}-\dot{\boldsymbol{p}}_{h_a})+S_F(\boldsymbol{p}_{h_d}-\boldsymbol{p}_{h_a})+M_F\ddot{\boldsymbol{p}}_{h_d}}{M_F}, \tag{6.15}$$

$$\ddot{\boldsymbol{\Theta}}_{h_a}=\frac{\boldsymbol{\tau}_{h_a}+D_\tau(\dot{\boldsymbol{\Theta}}_{h_d}-\dot{\boldsymbol{\Theta}}_{h_a})+S_\tau(\boldsymbol{\Theta}_{h_d}-\boldsymbol{\Theta}_{h_a})+M_\tau\ddot{\boldsymbol{\Theta}}_{h_d}}{M_\tau}, \tag{6.16}$$

where M_F, $M_\tau \in \mathbb{R}^+$ are the virtual mass, the diagonal positive semidefinite constant matrices D_F, D_τ, S_F, $S_\tau \in \mathbb{R}^{3\times3}$ that define a Hurwitz polynomial are the damping and stiffness constants that are used to change the physical properties of the UAV for position and orientation. Note that the elements of D_F, D_τ, S_F and S_τ are ≥ 0. All these values can be chosen in order to provide a human friendly behavior avoiding sudden accelerations and allowing to exert forces on the UAV. In order to have a complete reference trajectory in the form $\boldsymbol{p}_{h_a}(t)$, $\dot{\boldsymbol{p}}_{h_a}(t)$, $\ddot{\boldsymbol{p}}_{h_a}(t)$, the values of $\dot{\boldsymbol{p}}_{h_a}$ and $\boldsymbol{p}_{h_a}$ are computed by integrating $\ddot{\boldsymbol{p}}_{h_a}$ in time. Similarly, $\dot{\boldsymbol{\Theta}}_{h_a}$ and $\boldsymbol{\Theta}_{h_a}$ are computed by integrating $\ddot{\boldsymbol{\Theta}}_{h_a}$.

6.3.3 Adaptive Super Twisting Control

The main objective of the low level controller is to compute the motor commands such that the 6DOF fully actuated multirotor is able to track the reference trajectory $\boldsymbol{p}_{h_a}$, $\dot{\boldsymbol{p}}_{h_a}$, $\ddot{\boldsymbol{p}}_{h_a}$, $\boldsymbol{\Theta}_{h_a}$, $\dot{\boldsymbol{\Theta}}_{h_a}$, $\ddot{\boldsymbol{\Theta}}_{h_a}$ provided by the admittance controller. Here, we propose our solution for the trajectory tracking problem of the multirotor in the presence of lumped disturbance $\boldsymbol{\kappa}_h$ by means of an adaptive super twisting controller (proposed earlier in Chapter. 4), a nonlinear method that has been proved efficient in presence of parameter uncertainties and unknown disturbances. Figure 6.4 shows the control scheme architecture of the developed controller, which includes several components as detailed in the rest of this Section.

Regular Control Form

The first step to derive the adaptive super twisting control equations is to simplify the multirotor dynamic model into a form that can be easily used for control design. Using the similar philosophy as in Chapter. 4, the dynamic model (6.6) can be written in state-space form:

$$\dot{\boldsymbol{x}}_h = \bar{\boldsymbol{f}}_h(\boldsymbol{x}) + \bar{\boldsymbol{g}}_h(\boldsymbol{x})\boldsymbol{u}_h \tag{6.17}$$

where

$$\boldsymbol{x}_h = \begin{bmatrix} x_h & y_h & z_h & \phi_h & \theta_h & \psi_h & \dot{x}_h & \dot{y}_h & \dot{z}_h & p_h & q_h & r_h \end{bmatrix}^T \in \mathbb{R}^{12\times1} \tag{6.18}$$

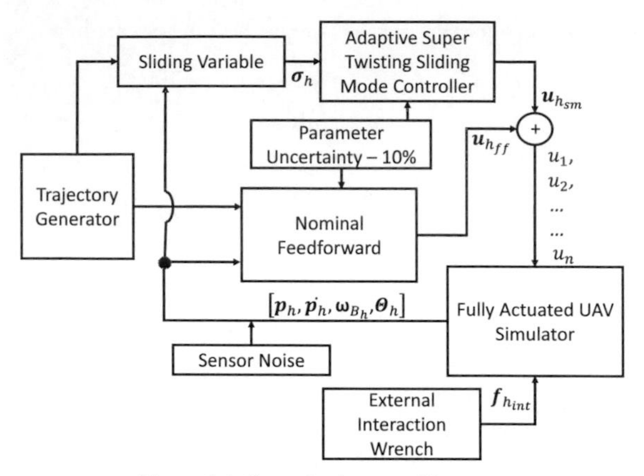

Figure 6.4: Control scheme architecture.

$$\bar{\boldsymbol{f}}_h(\boldsymbol{x}) = \begin{bmatrix} \begin{bmatrix} \dot{x}_h \\ \dot{y}_h \\ \dot{z}_h \\ \bar{f}_{h_{(4,1)}} \\ \bar{f}_{h_{(5,1)}} \\ \bar{f}_{h_{(6,1)}} \end{bmatrix} \\ \\ [\boldsymbol{f}_h]_{6\times 1} \end{bmatrix}, \bar{\boldsymbol{g}}_h(\boldsymbol{x}) = \begin{bmatrix} [\boldsymbol{0}_{6\times n}] \\ \\ [\boldsymbol{J}(\boldsymbol{\alpha}_h, \boldsymbol{\beta}_h, \boldsymbol{\lambda}_h, \boldsymbol{L}_{x_h})]_{6\times n} \end{bmatrix} \tag{6.19}$$

$$\boldsymbol{u} = \boldsymbol{u}_h = \begin{bmatrix} u_1 \\ u_2 \\ \vdots \\ u_n \end{bmatrix} = \begin{bmatrix} \bar{\omega}_{h1}^2 \\ \bar{\omega}_{h2}^2 \\ \vdots \\ \bar{\omega}_{hn}^2 \end{bmatrix}_{n\times 1} \tag{6.20}$$

with

$$\begin{cases} \bar{f}_{h_{(4,1)}} = p_h + q_h \sin\phi_h \tan\theta_h + r_h \cos\phi_h \tan\theta_h \\ \bar{f}_{h_{(5,1)}} = q_h \cos\phi_h - r_h \cos\phi_h \\ \bar{f}_{h_{(6,1)}} = q_h \sin\phi_h \sec\theta_h + r_h \cos\phi_h \sec\theta_h \end{cases}$$

In order to ensure that the propeller speed is always feasible, we take the following assumption on the control input (6.20):

Assumption 6.1:

The control input is bounded, i.e., $\boldsymbol{u}_h \in \mathcal{U}_h = \{\boldsymbol{u}_h^\star \in [\boldsymbol{u}_{h_{min}}, \boldsymbol{u}_{h_{max}}]\}$.

It is well known that a fully actuated multirotor is dynamically feedback linearizable with output

$$\boldsymbol{y}_h = \bar{\boldsymbol{h}}_h(\boldsymbol{x}) = \begin{bmatrix} x_h & y_h & z_h & \phi_h & \theta_h & \psi_h \end{bmatrix}^T. \tag{6.21}$$

Therefore, there exists a diffeomorphism $\Phi_h(\bar{\boldsymbol{x}})$ such that the coordinates transformation $\boldsymbol{z}_h = \Phi_h(\bar{\boldsymbol{x}})$ defined by

$$\begin{cases} z_{h_1} = x_h, & z_{h_2} = \dot{x}_h, & z_{h_3} = y_h, & z_{h_4} = \dot{y}_h, \\ z_{h_5} = z_h, & z_{h_6} = \dot{z}_h, & z_{h_7} = \phi_h, & z_{h_8} = p_h, \\ z_{h_9} = \theta_h, & z_{h_{10}} = q_h, & z_{h_{11}} = \psi_h, & z_{h_{12}} = r_h \end{cases} \tag{6.22}$$

transforms (6.17) into a regular form in which the dynamics of the output $\boldsymbol{y}_h$ in (6.21) are decoupled into a chain of integrators. The system transformation with the new states $\boldsymbol{z}_h = [z_{h_1}, z_{h_2}, \ldots\ldots, z_{h_{12}}]^T$ can be written in state-space form as

$$\dot{\boldsymbol{z}} = \begin{bmatrix} z_{h_2} \\ a_{h_x}(z) \\ z_{h_4} \\ a_{h_y}(z) \\ z_{h_6} \\ a_{h_z}(z) \\ z_{h_8} \\ a_{h_\phi}(z) \\ z_{h_{10}} \\ a_{h_\theta}(z) \\ z_{h_{12}} \\ a_{h_\psi}(z) \end{bmatrix} + \begin{bmatrix} \boldsymbol{0}_{1\times n} \\ b_{h_x}(z) \\ \boldsymbol{0}_{1\times n} \\ b_{h_y}(z) \\ \boldsymbol{0}_{1\times n} \\ b_{h_z}(z) \\ \boldsymbol{0}_{1\times n} \\ b_{h_\phi}(z) \\ \boldsymbol{0}_{1\times n} \\ b_{h_\theta}(z) \\ \boldsymbol{0}_{1\times n} \\ b_{h_\psi}(z) \end{bmatrix} \begin{bmatrix} u_1 \\ u_2 \\ \vdots \\ u_n \end{bmatrix}, \tag{6.23}$$

where $\boldsymbol{a}_h(\boldsymbol{z}) = \boldsymbol{f}_h$ and $\boldsymbol{b}_h(\boldsymbol{z}) = \boldsymbol{J}(\boldsymbol{\alpha}_h, \boldsymbol{\beta}_h, \boldsymbol{\lambda}_h, \boldsymbol{L}_{x_h})$. Therefore, the system model expressed in $\boldsymbol{z}_h$ and $\boldsymbol{u}_h$ is

$$\begin{bmatrix} \ddot{\boldsymbol{p}}_h \\ \ddot{\boldsymbol{\Theta}}_h \end{bmatrix} = \begin{bmatrix} \ddot{x}_h \\ \ddot{y}_h \\ \ddot{z}_h \\ \ddot{\phi}_h \\ \ddot{\theta}_h \\ \ddot{\psi}_h \end{bmatrix} = \begin{bmatrix} \dot{z}_{h_2} \\ \dot{z}_{h_4} \\ \dot{z}_{h_6} \\ \dot{z}_{h_8} \\ \dot{z}_{h_{10}} \\ \dot{z}_{h_{12}} \end{bmatrix} = \underbrace{\begin{bmatrix} a_{h_x}(z) \\ a_{h_y}(z) \\ a_{h_z}(z) \\ a_{h_\phi}(z) \\ a_{h_\theta}(z) \\ a_{h_\psi}(z) \end{bmatrix}}_{\triangleq \boldsymbol{a}_h(\boldsymbol{z})} + \underbrace{\begin{bmatrix} b_{h_x}(z) \\ b_{h_y}(z) \\ b_{h_z}(z) \\ b_{h_\phi}(z) \\ b_{h_\theta}(z) \\ b_{h_\psi}(z) \end{bmatrix}}_{\triangleq \boldsymbol{b}_h(\boldsymbol{z})} \boldsymbol{u}_h. \tag{6.24}$$

In order to ensure that the matrix $\boldsymbol{b}_h(z)$ in (6.24) is nonsingular, we take the following assumption

Assumption 6.2:
The roll and pitch angles ϕ_h and θ_h are limited to $(-\pi/2,\ \pi/2)$.

In fact, Assumption 6.2 ensures that ${}^{W}\boldsymbol{R}_{B_h}$ in (6.5) is nonsingular and invertible with $rank(\boldsymbol{b}_h(z)) = 6$.

Uncertainties

The model (6.24) describes the system without uncertainties. In order to incorporate the effect of disturbances and inexact knowledge of the parameters, we assume that:

1. the multirotor is subject to external disturbances $\boldsymbol{\chi}_h$ that acting on the CoM as force and torque wrenches. The dynamic model (6.24) becomes

$$\begin{bmatrix} \ddot{\boldsymbol{p}}_h \\ \ddot{\boldsymbol{\Theta}}_h \end{bmatrix} = \boldsymbol{a}_h(z) + \boldsymbol{b}_h(z)(\boldsymbol{u}_h + \boldsymbol{\chi}_h); \tag{6.25}$$

2. only the dynamic parameters m_h, $\boldsymbol{I}_{B_h}$ are uncertain.

Under this assumption, (6.25) becomes

$$\begin{aligned} \begin{bmatrix} \ddot{\boldsymbol{p}}_h \\ \ddot{\boldsymbol{\Theta}}_h \end{bmatrix} &= \boldsymbol{a}_{h_n} + \Delta\boldsymbol{a}_h + \boldsymbol{b}_{h_n}(\boldsymbol{u}_h + \boldsymbol{\chi}_h) + \Delta\boldsymbol{b}_h(\boldsymbol{u}_h + \boldsymbol{\chi}_h) = \\ &= \boldsymbol{a}_{h_n} + \boldsymbol{b}_{h_n}\boldsymbol{u}_h + \boldsymbol{\kappa}_h\ , \end{aligned} \tag{6.26}$$

where

- $\boldsymbol{a}_{h_n}$ and $\boldsymbol{b}_{h_n}$ describe the nominal model of the robot;
- $\Delta\boldsymbol{a}_h$ and $\Delta\boldsymbol{b}_h$ represent the parametric uncertainties;
- $\boldsymbol{\kappa}_h = \boldsymbol{b}_{h_n}\boldsymbol{\chi}_h + \Delta\boldsymbol{a}_h + \Delta\boldsymbol{b}_h(\boldsymbol{u}_h + \boldsymbol{\chi}_h)$ is the vector of lumped perturbations.

Note that $\boldsymbol{b}_{h_n}$ is always full rank (Assumption 6.2), so the lumped perturbations satisfy the matching condition. In general, it is possible to assume that the lumped perturbations are bounded. However, in practice it is difficult to estimate the upper bound on $\boldsymbol{\kappa}_h$. Hence, we take the following assumption:

Assumption 6.3:
$\boldsymbol{\kappa}_h$ is bounded as $\|\boldsymbol{\kappa}_h\|_2 \leq \kappa_{h_{\max}}$, but the bound $\kappa_{h_{\max}} \geq 0$ is unknown.

Control Law

The tracking controller is designed as a robust law $\boldsymbol{u}_h$ of the form

$$\boldsymbol{u}_h = \boldsymbol{u}_{h_{sm}} + \boldsymbol{u}_{h_{ff}}, \tag{6.27}$$

where

- $\boldsymbol{u}_{h_{sm}}$ is a term based on the sliding mode approach;
- $\boldsymbol{u}_{h_{ff}}$ is a feedforward term based on the dynamic inversion of the nominal model.

The sliding mode control term $\boldsymbol{u}_{h_{sm}}$ is designed to steer to zero the tracking errors of position $\boldsymbol{e}_{h_p} = \boldsymbol{p}_h - \boldsymbol{p}_{h_a} = \begin{bmatrix} e_{h_x} & e_{h_y} & e_{h_z} \end{bmatrix}^T \in \mathbb{R}^3$ and orientation $\boldsymbol{e}_{h_\Theta} = \boldsymbol{\Theta}_h - \boldsymbol{\Theta}_{h_a} = \begin{bmatrix} e_{h_\phi} & e_{h_\theta} & e_{h_\psi} \end{bmatrix}^T \in \mathbb{R}^3$ in presence of the uncertainties $\boldsymbol{\kappa}_h$. As explained in Sec. 6.3.3, the output is decoupled in the model in regular form (6.24). Assuming that $\boldsymbol{R}_{h_a}(t) \in \bar{\mathcal{C}}^3$ and $\boldsymbol{\omega}_{h_a} = [\boldsymbol{R}_{h_a}^T \dot{\boldsymbol{R}}_{h_a}]_\vee$, where $[\cdot]_\vee$ represents the inverse (vee) operator from $so(3)$ to $\mathbb{R}^3$, the attitude tracking error $\boldsymbol{e}_{h_R} \in \mathbb{R}^3$ is defined similarly to Lee *et al.* (2010) as

$$\boldsymbol{e}_{h_R} = \frac{1}{2}[\boldsymbol{R}_{h_a}^T\,{}^W\boldsymbol{R}_{B_h} - {}^W\boldsymbol{R}_{B_h}^T \boldsymbol{R}_{h_a}]_\vee, \tag{6.28}$$

and the tracking error of the angular velocity $\boldsymbol{e}_{h_\omega} \in \mathbb{R}^3$ is given by

$$\boldsymbol{e}_{h_\omega} = \boldsymbol{\omega}_{B_h} - {}^W\boldsymbol{R}_{B_h}^T \boldsymbol{R}_{h_a} \boldsymbol{\omega}_{h_a}. \tag{6.29}$$

Therefore, we chose the sliding variable as

$$\boldsymbol{\sigma}_h = \begin{bmatrix} \sigma_{h_x} \\ \sigma_{h_y} \\ \sigma_{h_z} \\ \sigma_{h_\phi} \\ \sigma_{h_\theta} \\ \sigma_{h_\psi} \end{bmatrix} = \begin{bmatrix} \dot{e}_{h_x} + \lambda_{x_1} e_{h_x} \\ \dot{e}_{h_y} + \lambda_{y_1} e_{h_y} \\ \dot{e}_{h_z} + \lambda_{z_1} e_{h_z} \\ \boldsymbol{e}_{h_\omega}(1) + \lambda_{\phi_1} \boldsymbol{e}_{h_R}(1) \\ \boldsymbol{e}_{h_\omega}(2) + \lambda_{\theta_1} \boldsymbol{e}_{h_R}(2) \\ \boldsymbol{e}_{h_\omega}(3) + \lambda_{\psi_1} \boldsymbol{e}_{h_R}(3) \end{bmatrix}, \tag{6.30}$$

where $\lambda_h \in \mathbb{R}^{n \times n}$ is a positive definite diagonal matrix. From (6.24), the time derivative of $\boldsymbol{\sigma}_h$ is

$$\dot{\boldsymbol{\sigma}}_h = \begin{bmatrix} -\ddot{x}_{h_a} + \lambda_{x_1} \dot{e}_{h_x} \\ -\ddot{y}_{h_a} + \lambda_{y_1} \dot{e}_{h_y} \\ -\ddot{z}_{h_a} + \lambda_{z_1} \dot{e}_{h_z} \\ -\dot{\boldsymbol{\omega}}_{h_a}(1) + \lambda_{\phi_1} \dot{\boldsymbol{e}}_{h_R}(1) \\ -\dot{\boldsymbol{\omega}}_{h_a}(2) + \lambda_{\theta_1} \dot{\boldsymbol{e}}_{h_R}(2) \\ -\dot{\boldsymbol{\omega}}_{h_a}(3) + \lambda_{\psi_1} \dot{\boldsymbol{e}}_{h_R}(3) \end{bmatrix} + \boldsymbol{a}_h(z) + \boldsymbol{b}_h(z)\,\boldsymbol{u}_h \tag{6.31}$$

hence $\boldsymbol{\sigma}_h$ has relative degree one with respect to $\boldsymbol{u}_h$. To achieve the 2-sliding mode $\boldsymbol{\sigma}_h = \dot{\boldsymbol{\sigma}}_h = \mathbf{0}$, we implement $\boldsymbol{u}_{h_{sm}}$ according to the Super Twisting controller (STC) Shtessel *et al.* (2014); Levant (1993). The equations of the standard STC are

$$\begin{aligned} \boldsymbol{u}_{h_{sm}} &= \boldsymbol{b}_h(z)^{-1}\left(-\bar{\boldsymbol{\alpha}}_h |\boldsymbol{\sigma}_h|^{\frac{1}{2}} sign(\boldsymbol{\sigma}_h) + \boldsymbol{v}_h\right) \\ \dot{\boldsymbol{v}}_h &= \begin{cases} -\boldsymbol{u}_{h_{sm}} & \text{if} |\boldsymbol{u}_{h_{sm}}| > \boldsymbol{u}_{h_m} \\ -\bar{\beta}_h sign(\boldsymbol{\sigma}_h) & \text{if} |\boldsymbol{u}_{h_{sm}}| \leq \boldsymbol{u}_{h_m} \end{cases}, \end{aligned} \tag{6.32}$$

where $\bar{\boldsymbol{\alpha}}_h$, $\bar{\boldsymbol{\beta}}_h$ are definite positive diagonal gain matrices and $\boldsymbol{u}_{h_m}$ denotes an upper bound for $\boldsymbol{u}_{h_{sm}}$. The control law (6.32) has two important properties, i) it does not require the knowledge of $\dot{\boldsymbol{\sigma}}_h$ and therefore of the linear acceleration $\ddot{\boldsymbol{p}}_h$ and angular acceleration $\ddot{\boldsymbol{\Theta}}_h$, and ii) the $sign(\boldsymbol{\sigma}_h)$, which represent a discontinuity, is integrated, thus significantly attenuating chattering.

In Shtessel *et al.* (2014) it is proved that the standard STC achieves finite-time convergence to the 2^{nd}order-sliding manifold with few assumptions. In particular, it is necessary to choose the gains $\bar{\boldsymbol{\alpha}}_h$ and $\bar{\boldsymbol{\beta}}_h$ high enough, according to the upper bound on $\boldsymbol{\kappa}_h$. However, being the upper bound on $\boldsymbol{\kappa}_h$ unknown (Assumption 6.3), a common problem in such situation is an over-conservative gain tuning which leads to unnecessary high control actions, chattering and noise amplification (see e.g., Rajappa *et al.* (2016)). In order to avoid these undesirable behaviors, it is possible to implement an adaptation law to select the gains online. Here we follow the law proposed in Shtessel *et al.* (2010, 2012),

$$\begin{aligned} \dot{\bar{\boldsymbol{\alpha}}}_h &= \begin{cases} \boldsymbol{\omega}_{h\alpha}\sqrt{\frac{\boldsymbol{\gamma}_h}{2}} sign\left(|\boldsymbol{\sigma}_h| - \boldsymbol{\mu}_h\right), & \text{if } \bar{\boldsymbol{\alpha}}_h > \bar{\boldsymbol{\alpha}}_{h_m} \\ \boldsymbol{\eta}_h, & \text{if } \bar{\boldsymbol{\alpha}}_h \leq \bar{\boldsymbol{\alpha}}_{h_m} \end{cases} \\ \bar{\boldsymbol{\beta}}_h &= 2\varepsilon_h \boldsymbol{\alpha}_h , \end{aligned} \tag{6.33}$$

where

- $\bar{\boldsymbol{\alpha}}_{h_m}$ is an arbitrary small positive constant introduced to keep the gains positive;
- $\boldsymbol{\mu}_h$ is a positive parameter defining the boundary layer for the real sliding mode.
- $\boldsymbol{\omega}_{h_\alpha}, \boldsymbol{\gamma}_h, \boldsymbol{\eta}_h$ are arbitrary positive constants;

Under few assumptions Shtessel *et al.* (2012), the STC with adaptive gains (6.33) achieves finite-time convergence to a real 2-sliding mode $\|\boldsymbol{\sigma}_h\| \leq \boldsymbol{\mu}_{h_1}$ and $\|\boldsymbol{\sigma}_h\| \leq \boldsymbol{\mu}_{h_2}$, with $\boldsymbol{\mu}_{h_1} \geq \boldsymbol{\mu}_h$ and $\boldsymbol{\mu}_{h_2} \geq \mathbf{0}$. Note that, in order to achieve convergence, $\boldsymbol{\mu}_h$ in (6.33) must be selected properly. An incorrect value assignment could lead to either instability and the control gains shooting up to infinity or to poor accuracy Plestan *et al.* (2010). Here, according to Plestan *et al.* (2010), we select $\boldsymbol{\mu}_h$ as the time-varying function:

$$\boldsymbol{\mu}_h(t) = 4\,\bar{\boldsymbol{\alpha}}(t)\,T_{h_e}\,, \tag{6.34}$$

where T_{h_e} is the sampling time of the controller.

Note that, with the introduction of the gain adaptation law (6.33), the controller does not require any a priori knowledge of the upper bound of the lumped disturbance $\boldsymbol{\kappa}_h$. In fact, the gains $\bar{\boldsymbol{\alpha}}_h$ and $\bar{\boldsymbol{\beta}}_h$ are not chosen according to a worst case uncertainty, but rather they are increased only when necessary. The effect is a reduction of the chattering with respect to the standard STC.

Feedforward Control

The feedforward component $\boldsymbol{u}_{h_{ff}}$ in (6.27) is the wrench that needs to be applied to the nominal model of the UAV to track a reference trajectory in the absence of initial error. Its effect is to decrease the magnitude of the sliding mode control $\boldsymbol{u}_{h_{sm}}$, thus helping in reducing the gains of the ASTC and further attenuating chattering.

The expression of $\boldsymbol{u}_{h_{ff}}$ is obtained by dynamic inversion of the system model (6.24) as

$$\boldsymbol{u}_{h_{ff}} = \boldsymbol{b}_h(\boldsymbol{z})^{-1}\left(\begin{bmatrix}\ddot{x}_{h_d}\\ \ddot{y}_{h_d}\\ \ddot{z}_{h_d}\\ \ddot{\phi}_{h_d}\\ \ddot{\theta}_{h_d}\\ \ddot{\psi}_{h_d}\end{bmatrix} - \boldsymbol{a}_h(\boldsymbol{z})\right)\,. \tag{6.35}$$

6.4 Simulations and Analysis

The main objective of this validation is to study the behavior of the Human-UAV physical interaction with a fully actuated multirotor system. The simulations are performed in Matlab using the mathematical model of the hexarotor with tilted propellers presented in Chapter. 5 (Rajappa *et al.*, 2015) (hence $n = 6$). The parameters of the admittance controller are selected as: virtual mass M_F, $M_\tau = 1$, damping constant D_F, $D_\tau = 1$ and stiffness constant S_F, $S_\tau = 0$. The gain matrix in the observer is fixed at $\boldsymbol{K}_h = 5$. In addition, we provide a comparison between the case of a standard underactuated quadrotor and the fully actuated UAV to highlight the benefits of using a fully actuated UAV. All the experimental parameters are listed in Table. 6.1.

6.4.1 Human-Robot Physical Interaction with a Fully Actuated UAV

In the first simulation (Fig. 6.5 and Fig. 6.6), the fully actuated UAV, whose initial position and orientation are $[0\;\;0\;\;0]^T$ *m* and $[0\;\;0\;\;0]^T$ *rad* respectively, is first set in hovering

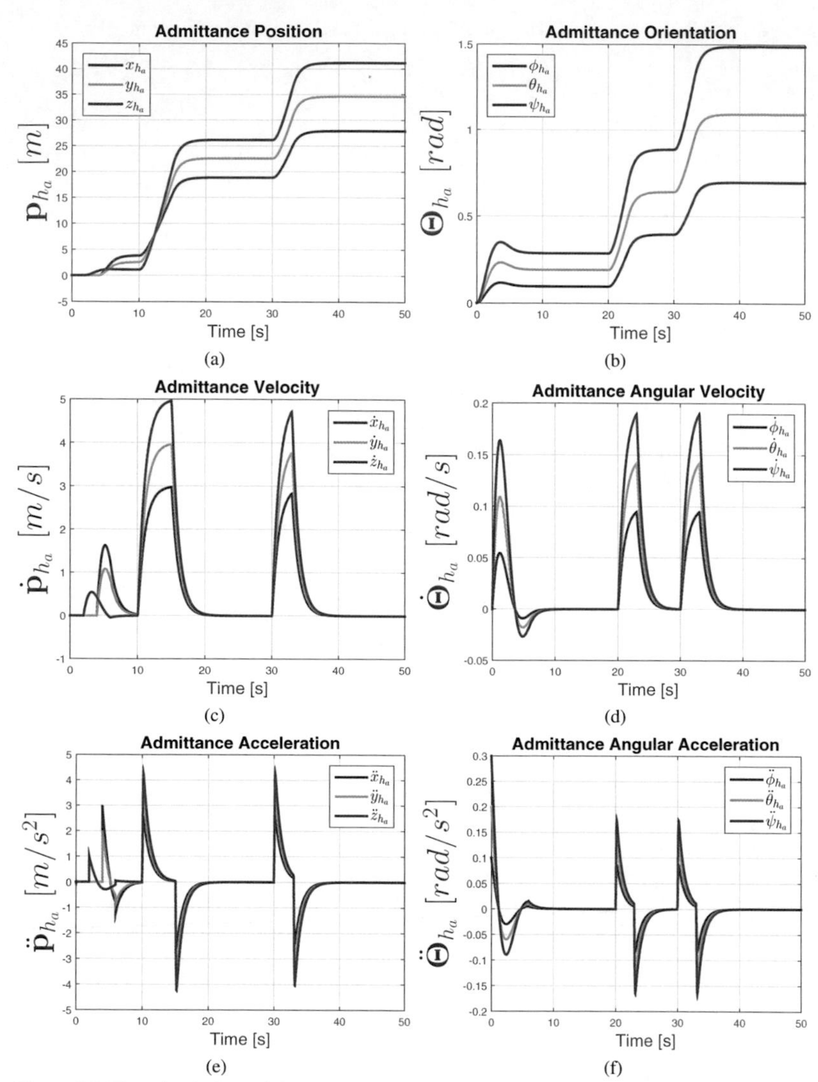

Figure 6.5: First simulation. 6.5(a,c,e): reference trajectory position, $\boldsymbol{p}_{h_a}$, velocity $\dot{\boldsymbol{p}}_{h_a}$, acceleration $\ddot{\boldsymbol{p}}_{h_a}$ respectively generated for x_{h_a}(red), y_{h_a}(green) and z_{h_a}(blue) by the admittance controller; 6.5(b,d,f): reference trajectory orientation, $\boldsymbol{\Theta}_{h_a}$, angular velocity $\dot{\boldsymbol{\Theta}}_{h_a}$, angular acceleration $\ddot{\boldsymbol{\Theta}}_{h_a}$ respectively generated for ϕ_{h_a}(red), θ_{h_a}(green) and ψ_{h_a}(blue) by the admittance controller.

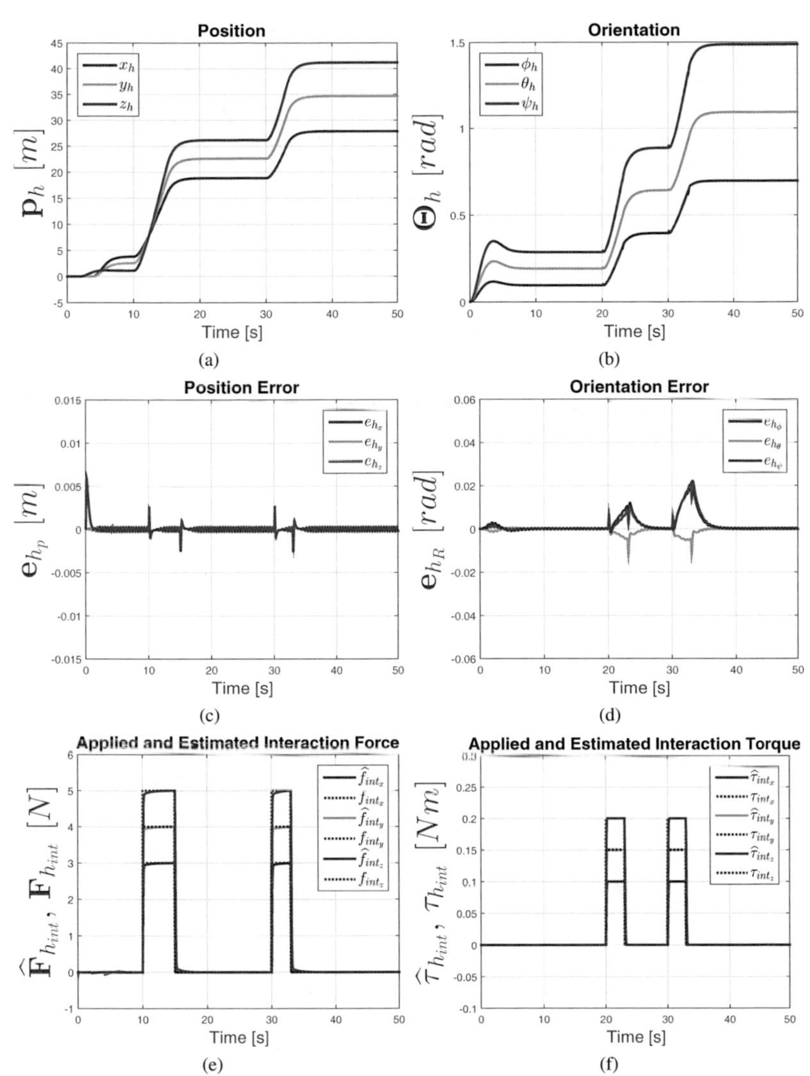

Figure 6.6: First simulation. 6.6(a) actual position $\boldsymbol{p}_h$ in x_h(red), y_h(green) and z_h(blue) of the UAV; 6.6(b) actual orientation $\boldsymbol{\Theta}_h$ in ϕ_h(red), θ_h(green) and ψ_h(blue) of the UAV; 6.6(c): position tracking error e_{h_p} in e_{h_x}(red), e_{h_y}(green) and e_{h_z}(blue). 6.6(d): orientation tracking error e_{h_R} in e_{h_ϕ}(red), e_{h_θ}(green) and e_{h_ψ}(blue); 6.6(e): applied (dashed) and estimated (solid) interaction force $\boldsymbol{F}_{h_{int}}$ with f_{int_x}(red), f_{int_y}(green) and f_{int_z}(blue); 6.6(f): applied (dashed) and estimated (solid) interaction torque $\boldsymbol{\tau}_{h_{int}}$ with τ_{int_x}(red), τ_{int_y}(green) and τ_{int_z}(blue).

at $z_h = 1\ m$. At $t = 2\ s$ a new hovering configuration is provided as $\boldsymbol{p}_h = [4\ \ 3\ \ 1]^T\ m$ and $\boldsymbol{\Theta}_h = [0.1\ \ 0.2\ \ 0.3]^T\ rad$. Between $t = 10\ s$ and $t = 15\ s$, an interaction force $\boldsymbol{F}_{h_{int}} = [3\ \ 4\ \ 5]^T\ N$ is applied as can be seen in Fig. 6.6(e) and the reference position $\boldsymbol{p}_{h_a}$ (Fig. 6.5(a)) is changed accordingly by the admittance controller. Note that during this phase the reference orientation $\boldsymbol{\Theta}_{h_a}$ (Fig. 6.5(b)) does not change. The corresponding reference linear velocities and accelerations and angular velocities and accelerations are shown in Fig. 6.5(c), Fig. 6.5(e), Fig. 6.5(d) and Fig. 6.5(f) respectively.

Similarly, between time $t = 20\ s$ and time $t = 23\ s$ an interaction torque with the strength of $\boldsymbol{\tau}_{h_{int}} = [0.1\ \ 0.15\ \ 0.2]^T\ Nm$ is applied (Fig. 6.6(f)), and the reference orientation $\boldsymbol{\Theta}_{h_a}$ changes accordingly whereas the reference position $\boldsymbol{p}_{h_a}$ remains unchanged (Fig. 6.6(a) and Fig. 6.6(b)). Finally, between time $t = 30\ s$ and $t = 33\ s$, both $\boldsymbol{F}_{h_{int}} = [3\ \ 4\ \ 5]^T\ N$ and $\boldsymbol{\tau}_{h_{int}} = [0.1\ \ 0.15\ \ 0.2]^T\ Nm$ are applied at the same time (Fig. 6.6(e), Fig. 6.6(f)). Consequently, as can be seen in Fig. 6.6(a) and Fig. 6.6(b), both the reference position and orientation changes accordingly to the interaction wrench. These plots demonstrates the proper operation of the admittance controller, and the smoothness of the reference trajectory.

Interaction Wrench Observer

For correct operation, the interaction wrench $\boldsymbol{\Lambda}_{h_{int}}$ must be properly estimated. In order to validate the Interaction Wrench Observer, Figures 6.6(e) and 6.6(f) report both the applied (dashed lines) and the estimated (solid lines) interaction wrenches. In both Figures it is possible to appreciate the high convergence speed of the estimates w.r.t. the real values. In particular, when the forces and/or torques are applied the estimates suddenly increases to more than 95% of the applied values, while the remaining error is reduced to negligible in less than $1\ s$. The convergence speed is important in order to guarantee a proper response of the UAV with respect to the interaction from the human, which results in a safer system.

ASTC

Similarly, once the reference trajectory is generated, in order to ensure the safety of the human interacting with the robot it is necessary to ensure that the UAV is able to properly follow such trajectory. Figures 6.6(c) and 6.6(d) report the plot of the position e_{h_p} and orientation e_{h_R} tracking error during the first simulation. In both plots, it is possible to appreciate how the errors remain very limited. Moreover, no chattering (which usually affects sliding mode controllers) is visible in the plots, thanks to the adaptation law implemented in order to automatically tune the controller online.

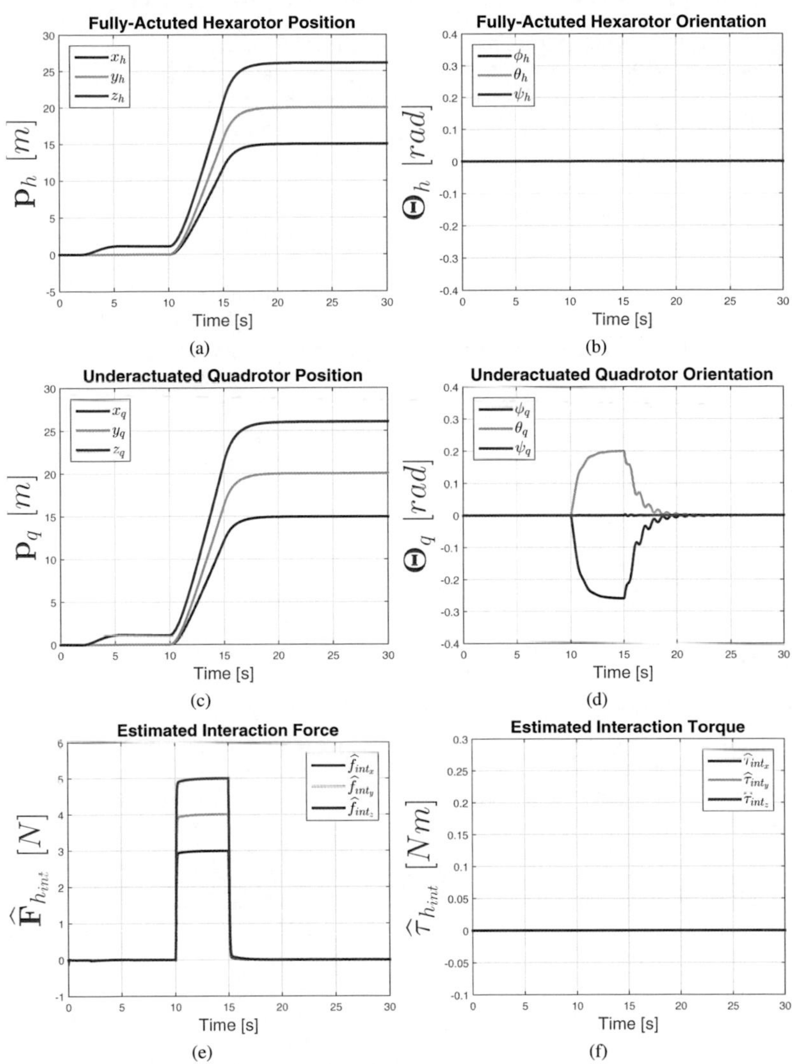

Figure 6.7: Results of the comparison between underactuated quadrotor and fully-actuated UAV. 6.7(a): actual position $\boldsymbol{p}_h$ of the fully actuated UAV with x_h(red), y_h(green) and z_h(blue); 6.7(b): actual orientation $\boldsymbol{\Theta}_h$ of the fully actuated UAV with roll(red), pitch(green) and yaw(blue); 6.7(c): current position $\boldsymbol{p}_q$ of the quadrotor with x_q(red), y_q(green) and z_q(blue); 6.7(d): current orientation $\boldsymbol{\Theta}_q$ of the quadrotor with roll(red), pitch(green) and yaw(blue).

6.4.2 Comparison of Underactuated and Fully-Actuated dynamics

In the second simulation (Fig. 6.7), a fully actuated UAV in hovering in $\boldsymbol{p}_h = [0\ 0\ 1]^T\ m$ and $\boldsymbol{\Theta}_h = [0\ 0\ 0]^T\ rad$ is subject, between $t = 10\ s$ and $t = 15\ s$, to an interaction force $\boldsymbol{F}_{h_{int}} = [3\ 4\ 5]^T\ N$ (Fig. 6.7(e)) and no interaction torque (Fig. 6.7(f)). The same simulation is performed with a standard model of an underactuated quadrotor. In order to produce a fair comparison, all the dynamical parameters, as well as the controller and estimator gains of the quadrotor were selected as the corresponding parameters and gains of the fully actuated UAV.

The plot of the actual position $\boldsymbol{p}_h$ (Fig. 6.7(a)) and orientation $\boldsymbol{\Theta}_h$ (Fig. 6.7(b)) of the fully actuated UAV shows that it is correctly responding to the interaction force by changing its position while keeping a still orientation. On the other hand, the same plots of the actual position $\boldsymbol{p}_q$ (Fig. 6.7(c)) and orientation $\boldsymbol{\Theta}_q$ (Fig. 6.7(d)) of the quadrotor show that the quadrotor changes its roll and pitch angles in order to respond to an interaction which encompasses only a force exchange. While this behavior is largely expected due to the underactuation of the quadrotor, its realization severely affects the safety and the fruibility of this system. In fact, the sudden change in orientation of the quadrotor needed to perform the lateral movement moves also the interaction surface firstly touched by the human, which could lose the grip (hence the contact) with the quadrotor. In addition, a disturbance in the orientation may render the quadrotor unstable.

6.5 Discussions and Possible Extensions

This chapter is the amalgamation of all the research work presented in the previous chapters on human-UAV physical interaction (Chapter. 3), robust controller (Chapter. 4) and fully actuated UAVs (Chapter. 5). Summarizing this chapter:

1. it is introduced the dynamic modeling (Sec. 6.2) of a generic non-coplanar fully actuated UAV which has propellers equal to or greater than 6. This way the methodologies developed can be used for any fully actuated UAV irrespective of the number of propellers.

2. it is developed a control architecture scheme and interaction wrench algorithm employing a fully actuated UAV (Sec. 6.3). This also highlights the underlying existing estimation and control problem that needs to be solved.

3. it is then implemented an admittance control to designate the reference trajectory which is obtained by not only the interaction forces but also the torques through human physical contact. Since a fully actuated UAV is used in this case, the UAV could orient itself in free space into a reference roll, pitch and yaw angle according to the control law provided in Sec. 6.3.2.

4. it is presented an adaptive super twisting controller (Sec. 6.3.3) that can be applied to a general $n-$rotor fully actuated UAV. This controller effectively rejects all the disturbances affecting the system, giving a robust performance while still tracking the reference trajectory provided by the admittance controller.

5. it is also presented the concept CAD model (Fig. 6.2) of one of the hardware architectures that could be designed to safely realize the concept of physical interaction using a fully actuated UAV.

6. it is provided the simulation comparison of the effectiveness of the system and its superior performance with respect to standard underactuated quadrotor in Sec. 6.4.2.

The work presented in this chapter could be extended is many possible ways. The most important will be the practical implementation of the proposed system. As highlighted earlier, since torque exchange is possible with this UAV platform, the effective applications will include a more friendly human-UAV physical interaction which would be safer than a quadorotor. There could be many improvements done in the interaction surface architecture, specifically designed taking safety as the primary concern. This could also include a sensor setup which includes all the interaction surface (e.g., tactile surface), which is computationally light and gives the exact location of point of contact. Further studies could be dedicated to more generalization of the concepts for any fully actuated UAV configuration which are not taken into consideration here. As mentioned earlier in the scientific direction, this platform can also be used for studies with respect to human subjects, to study their interaction behavior patterns with UAVs.

Parameter	Description	Value	Unit
m	mass of the UAV	1	Kg
g	gravity acceleration	9.81	m/s^2
I_{xx}	inertia along X-axis	0.011549	$Kg.m^2$
I_{yy}	inertia along Y-axis	0.011368	$Kg.m^2$
I_{zz}	inertia along Z-axis	0.019444	$Kg.m^2$
b	lift coefficient	$1.6073 * 10^{-5}$	N/Ω^2
d	drag coefficient	$2.7988 * 10^{-7}$	Nm/Ω^2
l	arm length	0.4	m
$\boldsymbol{\alpha}_h$	propeller tilt w.r.t. X-axis	0.698132	rad
$\boldsymbol{\beta}_h$	propeller tilt w.r.t. Y-axis	0.0872665	rad
$[\lambda_{x_1}, \lambda_{y_1}, \lambda_{z_1}]$	position error gain	[15, 15, 15]	-
$[\lambda_{\phi_1}, \lambda_{\theta_1}, \lambda_{\psi_1}]$	orientation error gain	[8, 8, 8]	-
$\boldsymbol{\omega}_{\boldsymbol{\alpha}_1}$	position constant	$[200, 200, 200] * 100$	-
$\boldsymbol{\omega}_{\boldsymbol{\alpha}_2}$	orientation constant	$[20, 20, 20] * 100$	-
$\boldsymbol{\gamma}$	positive constant	$[0.8, 0.8, 0.8, 0.8, 0.8, 0.8]$	-
$\boldsymbol{\alpha}_m$	minimum positive constant	0.1	-
$\boldsymbol{\eta}$	positive constant	$[0.1, 0.1, 0.1, 0.1, 0.1, 0.1]$	-
ε	positive constant	$diag[1, 1, 1, 1, 1, 1]$	-
T_e	sampling time	0.001	s
M_F	virtual mass constant-force	1	-
M_τ	virtual mass constant-torque	1	-
D_F	damping constant-force	1	-
D_τ	damping constant-torque	1	-
S_F	stiffness constant-force	0	-
S_τ	stiffness constant-torque	0	-
$\boldsymbol{K}_h = 5$	observer gain	5	-

Table 6.1: Experimental parameters of human-UAV physical interaction using fully actuated UAV

Chapter 7

Conclusions

In this dissertation *"Human-UAV Physical Interaction and Towards Fully Actuated Aerial vehicles"*, we have based our research study on two major topics namely: (i) Human-UAV physical interaction, and (ii) Fully actuated UAVs. Both separately have great potential to improve existing UAV research and to lead to new specific research direction for future investigations.

Initially, we started of extensively using *quadrotors*, the most common UAV research platform. Quadrotors are underactuated aerial vehicle, whose desired position and yaw could be designated at any particular time but not its desired roll and pitch angles. The quadrotor platform is detailed in Sec. 1.2 and the system dynamics are defined in Sec. 2.2.

Our very first objective was to develop a system to allow humans to physically interact with a UAV. Though there is not so much demand for such an interaction currently, the requirement for human and UAV share the same workspace is visible in many recent applications in aerial manipulation and interaction domains. Therefore the future holds the fact that *"Humans and UAVs would / should coexist"*. In order to realize such a revolutionary objective, it is important to have an exact knowledge of the interaction forces and torques (wrenches). More importantly, not only the information of external wrench is vital, but also the methodology of estimation should be feasible enough to be implemented in the existing UAV platform which has its known limitations in payload capacity, flight time and computational power.

Therefore in Chapter 2, we introduced and implemented a methodology for the *external wrench estimation* on the idea of residual-based generalized momenta. This technique doesn't require any additional force/ torque sensors to be installed thereby not affecting the flight time and is computationally very light. Utilizing the system states along with the generated motor commands input, all the external wrenches which are not part of the original UAV dynamics are gathered as residual at its center of mass. The residual later aids in the estimation of the external wrench (Sec. 2.3). Naturally, before utilizing it otherwise for interaction, the first temptation is to use it as a disturbance observer in existing state-of-the-art controllers for quadrotors. In line with this, we computed the roll and pitch disturbance compensation factor that could be then utilized as the feed-forward factor in the disturbance compensated near-hovering controller (Sec. 2.4). Note that this chapter also include in Sec. 2.6, the details of the hardware experimental setup specifying the electronic control boards, sensors, hardware and software framework for communication between different components.

Then, the external wrench observer has been employed as an observer in *Human-UAV physical interaction* (Chapter 3). Through the development of a button *sensor-ring* (Sec. 3.4), the external wrench was separated into disturbance and interaction wrenches (Sec. 3.5). This separation is obtained through the solution of an optimization problem (Sec. 3.5.2). With this method, we were also able to estimate the location of the point of contact (PoC), where the interaction occurred, so that appropriate counter action could be taken. With the separation of the disturbance and interaction wrenches in place, it is also equally important how the controller handles these wrenches. Therefore in Sec. 3.6, an *admittance control paradigm* (see Sec. 3.6.1) is developed which modifies the desired trajectory on the basis of the human interaction by modifying the physical properties of the UAV considering them as *mass-spring-damper* system. The estimated disturbances are instead rejected through a *modified geometric tracking* controller (Sec. 3.6.2) to track the reference trajectory provided by the admittance controller. This has been extensively validated both through hardware-in-the-loop simulations (Sec. 3.7) and experiments (Sec. 3.8) which proves that the human-UAV physical interaction is a feasible reality. During this work, we identified the main limitations in the proposed setup namely: (i) the interaction torques could not be exchanged with the UAVs because of the underactuated nature of the quadrotor UAV. Its dynamics could lead to instability and safety concerns, and (ii) the disturbance rejection must have a robust performance to be effective in situations of parameter uncertainties and external perturbations which are common during physical interaction scenarios.

In order to address these limitations, in Chapter 4, we proposed a robust *adaptive super twisting sliding mode* controller design for quadrotor UAVs. This approach comes under the non-linear controller category, where we derived the regular control form of the quadrotor UAV dynamics in the state space model. Because of the underactuated quadrotor dynamics, it was required to take the dynamic system states to an higher order up to snap ($4^{th} - degree$) in order to have a decoupled input-output relationship (Sec. 4.2.2). Important characteristics of the adaptive super twisting controller (Sec. 4.3) are: (i) compensation for all the uncertainties coming from parameters, modeling errors and disturbances, (ii) no requirement of the knowledge of the uncertainty bounds, (iii) works by the principle of gain adaptation thereby the control actions, chattering and noise amplification are reduced, and (iv) uses a feedforward dynamic inversion to reduce the discontinuous control. The adaptive behavior of the controller is also compared against the standard super twisting controller (Sec. 4.4.3) showing improved performances.

In Chapter 5, we addressed the other bigger limitation of under-actuation in UAV by proposing and designing a novel *tilted propeller hexarotor*. The hexarotor is fully-actuated with 6 DoF and therefore can position and orient itself in free space in any desired trajectory. Moreover, it could generate forces as well as torques in all the three axes. Sec. 5.2 and Sec. 5.3 details the translational / rotational modeling dynamics and the feedback linearization control architecture of the fully-actuated hexarotor setup respectively. The full-actuation could be attained at many different tilt angle configuration of the propellers. Therefore, we optimized the tilt angle based on the power consumption energy, so that the flight time of the UAV could be increased depending upon the task trajectory (Sec. 5.4.2). The architectural development of the hexarotor prototype is detailed in Sec. 5.6. The important features are: (i) the propeller tilting adaptor which could rotate in along the three axes and maintain the origin of the propeller frame (point where thrust is generated) always in the same position, and (ii) the direct velocity control of the brushless motor controller. This feature improves the prototype by directly passing the generated velocity control input to the low level controller without the requirement of any intermediate transformation.

Among the many advantages of using a fully-actuated UAV platform, one arises in human-UAV interaction tasks. The full-actuation allows admitting all the external disturbance forces and torques arising from the environment. During a physical interaction task this adds an additional safety compared to the standard quadrotor which cannot independently comply with a torque around the x and y axes not being able to assume arbitrary roll and pitch angles. This feature was further studied in Chapter 6 through the development of a fully actuated UAV model (Sec. 6.2) for human-UAV physical interaction (Sec. 6.3). In this scheme not only the methodologies of earlier discussion (Chapter. 3) were employed, but the robust adaptive super twisting controller scheme was adapted for this new UAV model (Sec. 6.3.3).

7.1 Future Research Directions

There are many possibilities to extend the work done in this dissertation. The discussion in all the previous chapters also highlighted the limitations that are present in methodology and technique used. Though some could be solved straightforwardly by addressing the problem for emerging research solutions, others were questioning the fundamental design in itself which requires new concepts (e.g. sensor setup, novel UAV designs). The following summary explores these limitations and the future research concepts.

1. The external wrench estimation technique (Chapter 2) can be employed as an easy to implement and efficient observer for disturbance rejection. Many of the controllers for aerial vehicles are ineffective when it comes to aggressive maneuvering application and unknown outdoor environment trajectory tracking. As the application domain is improving from indoor to outdoor autonomous navigation scenario, this observer could be employed without affecting the payload capacity. We have successfully implemented this technique in back-stepping based robust output regulation controller (Liu *et al.*, 2017a) for nontrivial quadrotor maneuvers and in nonlinear predictive controller for obstacle avoidance (Liu *et al.*, 2017b).

2. The observer gain parameter $\boldsymbol{K}_I$ (Chapter 2) tuning is essential for the estimation and convergence of wrench estimation. High gain could introduce noise which

adversely affects the controller performance whereas too low value results in poor convergence. This could be improved by introducing an adaptive behavior for the gain parameter depending on the system dynamics.

3. The hardware design (Chapter 3) proposed for human-UAV physical interaction could be further improved considering the safety of the interacting human. A full 3D structure could be a possibility in this case. Moreover, the current sensor setup fails in certain scenarios when the interaction and disturbance wrenches are in the same direction. This could be improved by a different sensor architecture.

4. Since human-UAV physical interaction is feasible, more application directions can be explored. For a quadrotor, it is already possible to react to an external yaw torque. This could be better explored for more interactive UAV behavior such as UAV rotating when it comes in contact with the vertical surface. Furthermore, many novel designs could be developed for UAV-HRPI with the focus on the application scenario.

5. The fully actuated hexarotor prototype (Chapter 5) could be used as the future UAV platform for aerial interaction and manipulation. Manipulators could be easily installed since it has better payload capability and its 6 DoF could be put into use during aerial manipulation. We have started to explore the possibilities in this direction.

6. The robustness of the fully actuated hexarotor could be further improved. This may be based on the tilt angles of the propellers and the tuning parameters of the controller. Further studies could be dedicated to the force and torque generated in a certain direction based on the tilt angles of the propellers. Its relationship with the power consumption also could be explored.

7. The proposed concept of human-UAV physical interaction with the fully actuated UAV (Chapter 6) could be realized. This would increase a variety of new application possibilities when the human could intuitively exchange not only forces but also torques with the UAV. We have started to work in this direction.

Appendix A

Technical Computations

A.1 Computation of Fully Actuated Hexarotor Model

In this section, we explain in detail the modeling of the tilted propeller hexarotor. The notations here follow similarly as defined in Chapter. 5 except with the subscript "h" being left out. Here the important hardware considerations that must be taken into notice are:

- The UAV considered here is a hexarotor. Therefore, the number of propellers are, $n = 6$.
- The origin of the ith propeller frame ($\boldsymbol{O}_{P_i}$) w.r.t body frame origin ($\boldsymbol{O}_B$) is given by

$$ {}^{B}\boldsymbol{p}_i = \boldsymbol{R}_Z(\lambda_i) \begin{bmatrix} L_{x_i} \\ 0 \\ 0 \end{bmatrix}, \quad \forall i = 1 \ldots 6 \tag{A.1} $$

$$ {}^{B}\boldsymbol{p}_i = \boldsymbol{R}_Z((i-1)\frac{2\pi}{n}) \begin{bmatrix} L_{x_i} \\ 0 \\ 0 \end{bmatrix}, \quad \forall i = 1 \ldots 6 \tag{A.2} $$

$$ {}^{B}\boldsymbol{p}_i = \begin{pmatrix} \cos((i-1)\frac{2\pi}{n}) & -\sin((i-1)\frac{2\pi}{n}) & 0 \\ \sin((i-1)\frac{2\pi}{n}) & \cos((i-1)\frac{2\pi}{n}) & 0 \\ 0 & 0 & 1 \end{pmatrix} \begin{pmatrix} L_{x_i} \\ 0 \\ 0 \end{pmatrix} \tag{A.3} $$

$$ {}^{B}\boldsymbol{p}_i = \begin{pmatrix} L_{x_i} \cos((i-1)\frac{2\pi}{n}) \\ L_{x_i} \sin((i-1)\frac{2\pi}{n}) \\ 0 \end{pmatrix} \tag{A.4} $$

  This means that all the propellers origins lie in the same plane as the hexarotor body origin.

- The robot architecture is symmetric with respect to its body center. Therefore the inertia matrix $\boldsymbol{I}_{B_h}$ of the hexacopter becomes

$$ \boldsymbol{I}_B = \begin{pmatrix} I_{xx} & 0 & 0 \\ 0 & I_{yy} & 0 \\ 0 & 0 & I_{zz} \end{pmatrix} \tag{A.5} $$

- Along with the symmetric configuration, the arm lengths are assumed to be of equal length

$$L_{x_1} = L_{x_2} = L_{x_3} = L_{x_4} = L_{x_5} = L_{x_6}; \tag{A.6}$$

The orientation of the ith propeller frame (F_{Pi}) w.r.t body frame (F_B) is given by

$${}^{B}\boldsymbol{R}_{P_i} = \boldsymbol{R}_Z((i-1)\frac{2\pi}{n})\boldsymbol{R}_X(\alpha_i)\boldsymbol{R}_Y(\beta_i), \quad \forall i = 1 \ldots 6 \tag{A.7}$$

$${}^{B}\boldsymbol{R}_{P_i} = \begin{pmatrix} \cos((i-1)\frac{2\pi}{n}) & -\sin((i-1)\frac{2\pi}{n}) & 0 \\ \sin((i-1)\frac{2\pi}{n}) & \cos((i-1)\frac{2\pi}{n}) & 0 \\ 0 & 0 & 1 \end{pmatrix} \begin{pmatrix} 1 & 0 & 0 \\ 0 & \cos(\alpha_i) & -\sin(\alpha_i) \\ 0 & \sin(\alpha_i) & \cos(\alpha_i) \end{pmatrix} \begin{pmatrix} \cos(\beta_i) & 0 & \sin(\beta_i) \\ 0 & 1 & 0 \\ -\sin(\beta_i) & 0 & \cos(\beta_i) \end{pmatrix} \tag{A.8}$$

$${}^{B}\boldsymbol{R}_{P_i} = \begin{pmatrix} c((i-1)\frac{2\pi}{n})c(\beta_i) - s((i-1)\frac{2\pi}{n})s(\alpha_i)s(\beta_i) & -s((i-1)\frac{2\pi}{n})c(\alpha_i) & c((i-1)\frac{2\pi}{n})s(\beta_i) + s((i-1)\frac{2\pi}{n})s(\alpha_i)c(\beta_i) \\ s((i-1)\frac{2\pi}{n})c(\beta_i) + c((i-1)\frac{2\pi}{n})s(\alpha_i)s(\beta_i) & c((i-1)\frac{2\pi}{n})c(\alpha_i) & s((i-1)\frac{2\pi}{n})s(\beta_i) - c((i-1)\frac{2\pi}{n})s(\alpha_i)c(\beta_i) \\ -c(\alpha_i)s(\beta_i) & s(\alpha_i) & c(\alpha_i)c(\beta_i) \end{pmatrix} \tag{A.9}$$

Here c($\star$)=cos($\star$) and s($\star$)= sin($\star$). The thrust is always generated along the Z-axis of the propeller frame and are related to the angular velocity of the rotating propellers which is given by,

$$\boldsymbol{T}_{Pi} = \begin{bmatrix} 0 & 0 & C_l \omega_i^2 \end{bmatrix}^T \tag{A.10}$$

Here C_l is the lift/thrust coefficient and $C_l > 0$

A.1.1 Translational Dynamics

From the Newton-Euler equations it is clear that the translational dynamics of the hexarotor system is defined by,

$$m \begin{pmatrix} \ddot{x} \\ \ddot{y} \\ \ddot{z} \end{pmatrix} = m \begin{pmatrix} 0 \\ 0 \\ -g \end{pmatrix} + \frac{1}{m} \left[{}^{W}R_B \sum_{i=1}^{n} {}^{B}\boldsymbol{R}_{Pi}.\boldsymbol{T}_{Pi} \right] \tag{A.11}$$

This equation can be written in parts separately for transforming all propeller thrusts into the body frame is given by

$$\sum_{i=1}^{6} {}^{B}R_{Pi}.\boldsymbol{T}_{Pi} = C_l \begin{pmatrix} c((i-1)\frac{2\pi}{n})s(\beta_i) + s((i-1)\frac{2\pi}{n})s(\alpha_i)c(\beta_i) & \ldots & \ldots & 6 \\ s((i-1)\frac{2\pi}{n})s(\beta_i) - c((i-1)\frac{2\pi}{n})s(\alpha_i)c(\beta_i) & \ldots & \ldots & 6 \\ c(\alpha_i)c(\beta_i) & \ldots & \ldots & 6 \end{pmatrix} \begin{pmatrix} \omega_1^2 \\ \ldots \\ \ldots \\ \omega_6^2 \end{pmatrix} \tag{A.12}$$

Utilizing the standard trigonometric values, the thrust generated in the individual propeller transformed to the body frame could be written as

Angle(deg)	sin θ	cos θ
0	0	1
60	$\frac{\sqrt{3}}{2}$	$\frac{1}{2}$
120	$\frac{\sqrt{3}}{2}$	$-\frac{1}{2}$
180	0	-1
240	$-\frac{\sqrt{3}}{2}$	$-\frac{1}{2}$
300	$-\frac{\sqrt{3}}{2}$	$\frac{1}{2}$

Propeller 1:

$$
{}^{B}R_{P1}.\boldsymbol{T}_{P1} = \begin{pmatrix} C_l\omega_1^2\left[c((1-1)\frac{2\pi}{6})s(\beta_1)+s((1-1)\frac{2\pi}{6})s(\alpha_1)c(\beta_1)\right] \\ C_l\omega_1^2\left[s((1-1)\frac{2\pi}{6})s(\beta_1)-c((1-1)\frac{2\pi}{6})s(\alpha_1)c(\beta_1)\right] \\ C_l\omega_1^2 c(\alpha_1)c(\beta_1) \end{pmatrix} = C_l\omega_1^2\begin{pmatrix} s(\beta_1) \\ -s(\alpha_1)c(\beta_1) \\ c(\alpha_1)c(\beta_1) \end{pmatrix} \tag{A.13}
$$

Propeller 2:

$$
{}^{B}R_{P2}.\boldsymbol{T}_{P2} = \begin{pmatrix} C_l\omega_2^2\left[c((2-1)\frac{2\pi}{6})s(\beta_2)+s((2-1)\frac{2\pi}{6})s(\alpha_2)c(\beta_2)\right] \\ C_l\omega_2^2\left[s((2-1)\frac{2\pi}{6})s(\beta_2)-c((2-1)\frac{2\pi}{6})s(\alpha_2)c(\beta_2)\right] \\ C_l\omega_2^2 c(\alpha_2)c(\beta_2) \end{pmatrix} = C_l\omega_2^2\begin{pmatrix} \frac{1}{2}s(\beta_2)+\frac{\sqrt{3}}{2}s(\alpha_2)c(\beta_2) \\ \frac{\sqrt{3}}{2}s(\beta_2)-\frac{1}{2}s(\alpha_2)c(\beta_2) \\ c(\alpha_2)c(\beta_2) \end{pmatrix} \tag{A.14}
$$

Propeller 3:

$$
{}^{B}R_{P3}.\boldsymbol{T}_{P3} = \begin{pmatrix} C_l\omega_3^2\left[c((3-1)\frac{2\pi}{6})s(\beta_3)+s((3-1)\frac{2\pi}{6})s(\alpha_3)c(\beta_3)\right] \\ C_l\omega_3^2\left[s((3-1)\frac{2\pi}{6})s(\beta_3)-c((3-1)\frac{2\pi}{6})s(\alpha_3)c(\beta_3)\right] \\ C_l\omega_3^2 c(\alpha_3)c(\beta_3) \end{pmatrix} = C_l\omega_3^2\begin{pmatrix} -\frac{1}{2}s(\beta_3)+\frac{\sqrt{3}}{2}s(\alpha_3)c(\beta_3) \\ \frac{\sqrt{3}}{2}s(\beta_3)+\frac{1}{2}s(\alpha_3)c(\beta_3) \\ c(\alpha_3)c(\beta_3) \end{pmatrix} \tag{A.15}
$$

Propeller 4:

$$
{}^{B}R_{P4}.\boldsymbol{T}_{P4} = \begin{pmatrix} C_l\omega_4^2\left[c((4-1)\frac{2\pi}{6})s(\beta_4)+s((4-1)\frac{2\pi}{6})s(\alpha_4)c(\beta_4)\right] \\ C_l\omega_4^2\left[s((4-1)\frac{2\pi}{6})s(\beta_4)-c((4-1)\frac{2\pi}{6})s(\alpha_4)c(\beta_4)\right] \\ C_l\omega_4^2 c(\alpha_4)c(\beta_4) \end{pmatrix} = C_l\omega_4^2\begin{pmatrix} -s(\beta_4) \\ s(\alpha_4)c(\beta_4) \\ c(\alpha_4)c(\beta_4) \end{pmatrix} \tag{A.16}
$$

Propeller 5:

$$
{}^{B}R_{P5}.\boldsymbol{T}_{P5} = \begin{pmatrix} C_l\omega_5^2\left[c((5-1)\frac{2\pi}{6})s(\beta_5)+s((5-1)\frac{2\pi}{6})s(\alpha_5)c(\beta_5)\right] \\ C_l\omega_5^2\left[s((5-1)\frac{2\pi}{6})s(\beta_5)-c((5-1)\frac{2\pi}{6})s(\alpha_5)c(\beta_5)\right] \\ C_l\omega_5^2 c(\alpha_5)c(\beta_5) \end{pmatrix} = C_l\omega_5^2\begin{pmatrix} -\frac{1}{2}s(\beta_5)-\frac{\sqrt{3}}{2}s(\alpha_5)c(\beta_5) \\ -\frac{\sqrt{3}}{2}s(\beta_5)+\frac{1}{2}s(\alpha_5)c(\beta_5) \\ c(\alpha_5)c(\beta_5) \end{pmatrix} \tag{A.17}
$$

Propeller 6:

$$
{}^{B}R_{P6}.\boldsymbol{T}_{P6} = \begin{pmatrix} C_l\omega_6^2\left[c((6-1)\frac{2\pi}{6})s(\beta_6)+s((6-1)\frac{2\pi}{6})s(\alpha_6)c(\beta_6)\right] \\ C_l\omega_6^2\left[s((6-1)\frac{2\pi}{6})s(\beta_6)-c((6-1)\frac{2\pi}{6})s(\alpha_6)c(\beta_6)\right] \\ C_l\omega_6^2 c(\alpha_6)c(\beta_6) \end{pmatrix} = C_l\omega_6^2\begin{pmatrix} \frac{1}{2}s(\beta_6)-\frac{\sqrt{3}}{2}s(\alpha_6)c(\beta_6) \\ -\frac{\sqrt{3}}{2}s(\beta_6)-\frac{1}{2}s(\alpha_6)c(\beta_6) \\ c(\alpha_6)c(\beta_6) \end{pmatrix} \tag{A.18}
$$

Therefore the total thrust is given as

$\sum_{i=1}^{6} {}^{B}R_{Pi}.\boldsymbol{T}_{Pi} =$

$$C_l \begin{pmatrix} s(\beta_1) & \frac{1}{2}s(\beta_2)+\frac{\sqrt{3}}{2}s(\alpha_2)c(\beta_2) & -\frac{1}{2}s(\beta_3)+\frac{\sqrt{3}}{2}s(\alpha_3)c(\beta_3) & -s(\beta_4) & -\frac{1}{2}s(\beta_5)-\frac{\sqrt{3}}{2}s(\alpha_5)c(\beta_5) & \frac{1}{2}s(\beta_6)-\frac{\sqrt{3}}{2}s(\alpha_6)c(\beta_6) \\ -s(\alpha_1)c(\beta_1) & \frac{\sqrt{3}}{2}s(\beta_2)-\frac{1}{2}s(\alpha_2)c(\beta_2) & \frac{\sqrt{3}}{2}s(\beta_3)+\frac{1}{2}s(\alpha_3)c(\beta_3) & s(\alpha_4)c(\beta_4) & -\frac{\sqrt{3}}{2}s(\beta_5)+\frac{1}{2}s(\alpha_5)c(\beta_5) & -\frac{\sqrt{3}}{2}s(\beta_6)-\frac{1}{2}s(\alpha_6)c(\beta_6) \\ c(\alpha_1)c(\beta_1) & c(\alpha_2)c(\beta_2) & c(\alpha_3)c(\beta_3) & c(\alpha_4)c(\beta_4) & c(\alpha_5)c(\beta_5) & c(\alpha_6)c(\beta_6) \end{pmatrix} \begin{pmatrix} \omega_1^2 \\ \omega_2^2 \\ \omega_3^2 \\ \omega_4^2 \\ \omega_5^2 \\ \omega_6^2 \end{pmatrix} \tag{A.19}$$

The transformation of the body frame to the world frame is given by the standard rotation matrix for the roll (ϕ), pitch (θ) and yaw (ψ) as

$${}^{W}R_{B} = \begin{pmatrix} \cos(\psi)\cos(\theta) & \cos(\psi)\sin(\theta)\sin(\phi)-\sin(\psi)\cos(\phi) & \cos(\psi)\sin(\theta)\cos(\phi)+\sin(\psi)\sin(\phi) \\ \sin(\psi)\cos(\theta) & \sin(\psi)\sin(\theta)\sin(\phi)+\cos(\psi)\cos(\phi) & \sin(\psi)\sin(\theta)\cos(\phi)-\cos(\psi)\sin(\phi) \\ -\sin(\theta) & \cos(\theta)\sin(\phi) & \cos(\theta)\cos(\phi) \end{pmatrix}. \tag{A.20}$$

Therefore transforming all the propeller thrust generated to the world frame as in (A.11), we have

$${}^{W}R_{B}\sum_{i=1}^{6} {}^{B}R_{Pi}.\boldsymbol{T}_{Pi} = \begin{pmatrix} \cos(\psi)\cos(\theta) & \cos(\psi)\sin(\theta)\sin(\phi)-\sin(\psi)\cos(\phi) & \cos(\psi)\sin(\theta)\cos(\phi)+\sin(\psi)\sin(\phi) \\ \sin(\psi)\cos(\theta) & \sin(\psi)\sin(\theta)\sin(\phi)+\cos(\psi)\cos(\phi) & \sin(\psi)\sin(\theta)\cos(\phi)-\cos(\psi)\sin(\phi) \\ -\sin(\theta) & \cos(\theta)\sin(\phi) & \cos(\theta)\cos(\phi) \end{pmatrix}$$

$$C_l \begin{pmatrix} s(\beta_1) & \frac{1}{2}s(\beta_2)+\frac{\sqrt{3}}{2}s(\alpha_2)c(\beta_2) & -\frac{1}{2}s(\beta_3)+\frac{\sqrt{3}}{2}s(\alpha_3)c(\beta_3) & -s(\beta_4) & -\frac{1}{2}s(\beta_5)-\frac{\sqrt{3}}{2}s(\alpha_5)c(\beta_5) & \frac{1}{2}s(\beta_6)-\frac{\sqrt{3}}{2}s(\alpha_6)c(\beta_6) \\ -s(\alpha_1)c(\beta_1) & \frac{\sqrt{3}}{2}s(\beta_2)-\frac{1}{2}s(\alpha_2)c(\beta_2) & \frac{\sqrt{3}}{2}s(\beta_3)+\frac{1}{2}s(\alpha_3)c(\beta_3) & s(\alpha_4)c(\beta_4) & -\frac{\sqrt{3}}{2}s(\beta_5)+\frac{1}{2}s(\alpha_5)c(\beta_5) & -\frac{\sqrt{3}}{2}s(\beta_6)-\frac{1}{2}s(\alpha_6)c(\beta_6) \\ c(\alpha_1)c(\beta_1) & c(\alpha_2)c(\beta_2) & c(\alpha_3)c(\beta_3) & c(\alpha_4)c(\beta_4) & c(\alpha_5)c(\beta_5) & c(\alpha_6)c(\beta_6) \end{pmatrix} \begin{pmatrix} \omega_1^2 \\ \omega_2^2 \\ \omega_3^2 \\ \omega_4^2 \\ \omega_5^2 \\ \omega_6^2 \end{pmatrix} \tag{A.21}$$

${}^{W}R_{B}\sum_{i=1}^{6} {}^{B}R_{Pi}.\boldsymbol{T}_{Pi} = C_l\, {}^{W}R_{B}$

$$\begin{pmatrix} s(\beta_1) & \frac{1}{2}s(\beta_2)+\frac{\sqrt{3}}{2}s(\alpha_2)c(\beta_2) & -\frac{1}{2}s(\beta_3)+\frac{\sqrt{3}}{2}s(\alpha_3)c(\beta_3) & -s(\beta_4) & -\frac{1}{2}s(\beta_5)-\frac{\sqrt{3}}{2}s(\alpha_5)c(\beta_5) & \frac{1}{2}s(\beta_6)-\frac{\sqrt{3}}{2}s(\alpha_6)c(\beta_6) \\ -s(\alpha_1)c(\beta_1) & \frac{\sqrt{3}}{2}s(\beta_2)-\frac{1}{2}s(\alpha_2)c(\beta_2) & \frac{\sqrt{3}}{2}s(\beta_3)+\frac{1}{2}s(\alpha_3)c(\beta_3) & s(\alpha_4)c(\beta_4) & -\frac{\sqrt{3}}{2}s(\beta_5)+\frac{1}{2}s(\alpha_5)c(\beta_5) & -\frac{\sqrt{3}}{2}s(\beta_6)-\frac{1}{2}s(\alpha_6)c(\beta_6) \\ c(\alpha_1)c(\beta_1) & c(\alpha_2)c(\beta_2) & c(\alpha_3)c(\beta_3) & c(\alpha_4)c(\beta_4) & c(\alpha_5)c(\beta_5) & c(\alpha_6)c(\beta_6) \end{pmatrix} \begin{pmatrix} \omega_1^2 \\ \omega_2^2 \\ \omega_3^2 \\ \omega_4^2 \\ \omega_5^2 \\ \omega_6^2 \end{pmatrix} \tag{A.22}$$

Substituting all the above derived values, the translation dynamics can be written as,

$$\begin{pmatrix} \ddot{x} \\ \ddot{y} \\ \ddot{z} \end{pmatrix} = \begin{pmatrix} 0 \\ 0 \\ -g \end{pmatrix} + \frac{C_l}{m}\, {}^{W}R_{B}$$

$$\begin{pmatrix} s(\beta_1) & \frac{1}{2}s(\beta_2)+\frac{\sqrt{3}}{2}s(\alpha_2)c(\beta_2) & -\frac{1}{2}s(\beta_3)+\frac{\sqrt{3}}{2}s(\alpha_3)c(\beta_3) & -s(\beta_4) & -\frac{1}{2}s(\beta_5)-\frac{\sqrt{3}}{2}s(\alpha_5)c(\beta_5) & \frac{1}{2}s(\beta_6)-\frac{\sqrt{3}}{2}s(\alpha_6)c(\beta_6) \\ -s(\alpha_1)c(\beta_1) & \frac{\sqrt{3}}{2}s(\beta_2)-\frac{1}{2}s(\alpha_2)c(\beta_2) & \frac{\sqrt{3}}{2}s(\beta_3)+\frac{1}{2}s(\alpha_3)c(\beta_3) & s(\alpha_4)c(\beta_4) & -\frac{\sqrt{3}}{2}s(\beta_5)+\frac{1}{2}s(\alpha_5)c(\beta_5) & -\frac{\sqrt{3}}{2}s(\beta_6)-\frac{1}{2}s(\alpha_6)c(\beta_6) \\ c(\alpha_1)c(\beta_1) & c(\alpha_2)c(\beta_2) & c(\alpha_3)c(\beta_3) & c(\alpha_4)c(\beta_4) & c(\alpha_5)c(\beta_5) & c(\alpha_6)c(\beta_6) \end{pmatrix} \begin{pmatrix} \omega_1^2 \\ \omega_2^2 \\ \omega_3^2 \\ \omega_4^2 \\ \omega_5^2 \\ \omega_6^2 \end{pmatrix} \tag{A.23}$$

This above equation (A.23) could be related with the translational dynamic equation (5.12) in Chapter. 5, where each individual component of $\boldsymbol{F}(\boldsymbol{\alpha}_h, \boldsymbol{\beta}_h, \boldsymbol{\lambda}_h)$ are derived.

A.1.2 Rotational Dynamics

The dynamic equation of the rotation dynamics is given by the Newton-Euler formulation

$$I_B \dot{\omega}_B = -\omega_B \times I_B \omega_B - \Sigma J_r (\omega \times e_3) \Omega_i + \tau_h. \tag{A.24}$$

The '$-\Sigma J_r (\omega \times e_3) \Omega_i$' term in the above equation can be neglected because we are using a brush-less motor set-up. Therefore the motor inertia can be neglected in the dynamics. Expanding the first term '$-\omega_B \times I_B \omega_B$', we get

$$-\omega_B \times I_B \omega_B = -\begin{pmatrix} p \\ q \\ r \end{pmatrix} \times \begin{pmatrix} I_{xx} & 0 & 0 \\ 0 & I_{yy} & 0 \\ 0 & 0 & I_{zz} \end{pmatrix} \begin{pmatrix} p \\ q \\ r \end{pmatrix} \tag{A.25}$$

$$-\omega_B \times I_B \omega_B = \begin{pmatrix} (I_{yy} - I_{zz}) qr \\ (I_{zz} - I_{xx}) pr \\ (I_{xx} - I_{yy}) pq \end{pmatrix}. \tag{A.26}$$

The main idea of our architecture is to fix α_i and β_i at a certain angle initially and not have any actuation for them during the flight. Therefore

$$\dot{\alpha}_i = \dot{\beta}_i = 0. \tag{A.27}$$

Hence there is no Gyroscopic moments and the adverse reaction moment that is generated because of the gyroscopic moments. The torques that are affecting the dynamics of the system are the propeller torques and the torque that are getting generated because of the thrust vector of the propellers. Now the torque generated on the body τ_h is given by,

$$\tau_h = \sum_{i=1}^{n} \left({}^B O_{Pi} \times {}^B R_{Pi}.T_{Pi} + {}^B R_{Pi}.T_{Di} \right). \tag{A.28}$$

Here it is noticed that the torque has two components. The component T_{Pi} is coming from the thrust generated in the ith propeller and T_{Di} is the drag component from the ith propeller. The drag component is given by the following equation

$$T_{Di} = \begin{bmatrix} 0 & 0 & (-1) C_d{}^i \omega_i^2 \end{bmatrix}^T \tag{A.29}$$

Here C_d is the drag coefficient and $C_d > 0$. The sign of C_d changes according to the propeller orientation setup. The origin of the ith propeller frame (O_{Pi}) w.r.t body frame origin (O_B) as derived previously is given as

$${}^B p_i = \begin{pmatrix} L_{x_i} \cos((i-1)\frac{2\pi}{n}) \\ L_{x_i} \sin((i-1)\frac{2\pi}{n}) \\ 0 \end{pmatrix} \tag{A.30}$$

Utilizing the trigonometric values, the components of individual propeller are written as

$${}^B O_{P1} = \begin{pmatrix} L_{x1} \\ 0 \\ 0 \end{pmatrix}; \quad {}^B O_{P2} = \begin{pmatrix} \frac{1}{2} L_{x2} \\ \frac{\sqrt{3}}{2} L_{x2} \\ 0 \end{pmatrix}; \quad {}^B O_{P3} = \begin{pmatrix} -\frac{1}{2} L_{x3} \\ \frac{\sqrt{3}}{2} L_{x3} \\ 0 \end{pmatrix}$$

$${}^B O_{P4} = \begin{pmatrix} -L_{x1} \\ 0 \\ 0 \end{pmatrix}; \quad {}^B O_{P5} = \begin{pmatrix} -\frac{1}{2} L_{x2} \\ -\frac{\sqrt{3}}{2} L_{x2} \\ 0 \end{pmatrix}; \quad {}^B O_{P6} = \begin{pmatrix} \frac{1}{2} L_{x3} \\ -\frac{\sqrt{3}}{2} L_{x3} \\ 0 \end{pmatrix}$$

Thrust Component:

The thrust component which affects the rotational dynamics can be written as

$$ {}^{B}R_{Pi}.\boldsymbol{T}_{Pi} = \begin{pmatrix} C_l\omega_i^2\left[c((i-1)\frac{2\pi}{n})s(\beta_i)+s((i-1)\frac{2\pi}{n})s(\alpha_i)c(\beta_i)\right] \\ C_l\omega_i^2\left[s((i-1)\frac{2\pi}{n})s(\beta_i)-c((i-1)\frac{2\pi}{n})s(\alpha_i)c(\beta_i)\right] \\ C_l\omega_i^2 c(\alpha_i)c(\beta_i) \end{pmatrix}. \tag{A.31} $$

For the individual propellers this can be written as,

Propeller 1:

$$ {}^{B}R_{P1}.T_{P1} = \begin{pmatrix} C_l\omega_1^2\left[c((1-1)\frac{2\pi}{6})s(\beta_1)+s((1-1)\frac{2\pi}{6})s(\alpha_1)c(\beta_1)\right] \\ C_l\omega_1^2\left[s((1-1)\frac{2\pi}{6})s(\beta_1)-c((1-1)\frac{2\pi}{6})s(\alpha_1)c(\beta_1)\right] \\ C_l\omega_1^2 c(\alpha_1)c(\beta_1) \end{pmatrix} = C_l\omega_1^2\begin{pmatrix} s(\beta_1) \\ -s(\alpha_1)c(\beta_1) \\ c(\alpha_1)c(\beta_1) \end{pmatrix} \tag{A.32} $$

Propeller 2:

$$ {}^{B}R_{P2}.T_{P2} = \begin{pmatrix} C_l\omega_2^2\left[c((2-1)\frac{2\pi}{6})s(\beta_2)+s((2-1)\frac{2\pi}{6})s(\alpha_2)c(\beta_2)\right] \\ C_l\omega_2^2\left[s((2-1)\frac{2\pi}{6})s(\beta_2)-c((2-1)\frac{2\pi}{6})s(\alpha_2)c(\beta_2)\right] \\ C_l\omega_2^2 c(\alpha_2)c(\beta_2) \end{pmatrix} = C_l\omega_2^2\begin{pmatrix} \frac{1}{2}s(\beta_2)+\frac{\sqrt{3}}{2}s(\alpha_2)c(\beta_2) \\ \frac{\sqrt{3}}{2}s(\beta_2)-\frac{1}{2}s(\alpha_2)c(\beta_2) \\ c(\alpha_2)c(\beta_2) \end{pmatrix} \tag{A.33} $$

Propeller 3:

$$ {}^{B}R_{P3}.T_{P3} = \begin{pmatrix} C_l\omega_3^2\left[c((3-1)\frac{2\pi}{6})s(\beta_3)+s((3-1)\frac{2\pi}{6})s(\alpha_3)c(\beta_3)\right] \\ C_l\omega_3^2\left[s((3-1)\frac{2\pi}{6})s(\beta_3)-c((3-1)\frac{2\pi}{6})s(\alpha_3)c(\beta_3)\right] \\ C_l\omega_3^2 c(\alpha_3)c(\beta_3) \end{pmatrix} = C_l\omega_3^2\begin{pmatrix} -\frac{1}{2}s(\beta_3)+\frac{\sqrt{3}}{2}s(\alpha_3)c(\beta_3) \\ \frac{\sqrt{3}}{2}s(\beta_3)+\frac{1}{2}s(\alpha_3)c(\beta_3) \\ c(\alpha_3)c(\beta_3) \end{pmatrix} \tag{A.34} $$

Propeller 4:

$$ {}^{B}R_{P4}.T_{P4} = \begin{pmatrix} C_l\omega_4^2\left[c((4-1)\frac{2\pi}{6})s(\beta_4)+s((4-1)\frac{2\pi}{6})s(\alpha_4)c(\beta_4)\right] \\ C_l\omega_4^2\left[s((4-1)\frac{2\pi}{6})s(\beta_4)-c((4-1)\frac{2\pi}{6})s(\alpha_4)c(\beta_4)\right] \\ C_l\omega_4^2 c(\alpha_4)c(\beta_4) \end{pmatrix} = C_l\omega_4^2\begin{pmatrix} -s(\beta_4) \\ s(\alpha_4)c(\beta_4) \\ c(\alpha_4)c(\beta_4) \end{pmatrix} \tag{A.35} $$

Propeller 5:

$$ {}^{B}R_{P5}.T_{P5} = \begin{pmatrix} C_l\omega_5^2\left[c((5-1)\frac{2\pi}{6})s(\beta_5)+s((5-1)\frac{2\pi}{6})s(\alpha_5)c(\beta_5)\right] \\ C_l\omega_5^2\left[s((5-1)\frac{2\pi}{6})s(\beta_5)-c((5-1)\frac{2\pi}{6})s(\alpha_5)c(\beta_5)\right] \\ C_l\omega_5^2 c(\alpha_5)c(\beta_5) \end{pmatrix} = C_l\omega_5^2\begin{pmatrix} -\frac{1}{2}s(\beta_5)-\frac{\sqrt{3}}{2}s(\alpha_5)c(\beta_5) \\ -\frac{\sqrt{3}}{2}s(\beta_5)+\frac{1}{2}s(\alpha_5)c(\beta_5) \\ c(\alpha_5)c(\beta_5) \end{pmatrix} \tag{A.36} $$

Propeller 6:

$$ {}^{B}R_{P6}.T_{P6} = \begin{pmatrix} C_l\omega_6^2\left[c((6-1)\frac{2\pi}{6})s(\beta_6)+s((6-1)\frac{2\pi}{6})s(\alpha_6)c(\beta_6)\right] \\ C_l\omega_6^2\left[s((6-1)\frac{2\pi}{6})s(\beta_6)-c((6-1)\frac{2\pi}{6})s(\alpha_6)c(\beta_6)\right] \\ C_l\omega_6^2 c(\alpha_6)c(\beta_6) \end{pmatrix} = C_l\omega_6^2\begin{pmatrix} \frac{1}{2}s(\beta_6)-\frac{\sqrt{3}}{2}s(\alpha_6)c(\beta_6) \\ -\frac{\sqrt{3}}{2}s(\beta_6)-\frac{1}{2}s(\alpha_6)c(\beta_6) \\ c(\alpha_6)c(\beta_6) \end{pmatrix} \tag{A.37} $$

Drag Component:

Similarly, the drag component which affects the rotational dynamics can be written as

$$ {}^{B}R_{Pi}.\boldsymbol{T}_{Di} = \begin{pmatrix} C_d\omega_i^2\left[c((i-1)\frac{2\pi}{n})s(\beta_i)+s((i-1)\frac{2\pi}{n})s(\alpha_i)c(\beta_i)\right] \\ C_d\omega_i^2\left[s((i-1)\frac{2\pi}{n})s(\beta_i)-c((i-1)\frac{2\pi}{n})s(\alpha_i)c(\beta_i)\right] \\ C_d\omega_i^2 c(\alpha_i)c(\beta_i) \end{pmatrix}. \tag{A.38} $$

For the individual propellers this can be written as,

Propeller 1:

$$
{}^{B}R_{P1}.T_{D1}=\begin{pmatrix} C_d\omega_1^2\left[c((1-1)\frac{2\pi}{6})s(\beta_1)+s((1-1)\frac{2\pi}{6})s(\alpha_1)c(\beta_1)\right] \\ C_d\omega_1^2\left[s((1-1)\frac{2\pi}{6})s(\beta_1)-c((1-1)\frac{2\pi}{6})s(\alpha_1)c(\beta_1)\right] \\ C_d\omega_1^2 c(\alpha_1)c(\beta_1) \end{pmatrix}=C_d\omega_1^2\begin{pmatrix} s(\beta_1) \\ -s(\alpha_1)c(\beta_1) \\ c(\alpha_1)c(\beta_1) \end{pmatrix} \tag{A.39}
$$

Propeller 2:

$$
{}^{B}R_{P2}.T_{D2}=\begin{pmatrix} C_d\omega_2^2\left[c((2-1)\frac{2\pi}{6})s(\beta_2)+s((2-1)\frac{2\pi}{6})s(\alpha_2)c(\beta_2)\right] \\ C_d\omega_2^2\left[s((2-1)\frac{2\pi}{6})s(\beta_2)-c((2-1)\frac{2\pi}{6})s(\alpha_2)c(\beta_2)\right] \\ C_d\omega_2^2 c(\alpha_2)c(\beta_2) \end{pmatrix}=C_d\omega_2^2\begin{pmatrix} \frac{1}{2}s(\beta_2)+\frac{\sqrt{3}}{2}s(\alpha_2)c(\beta_2) \\ \frac{\sqrt{3}}{2}s(\beta_2)-\frac{1}{2}s(\alpha_2)c(\beta_2) \\ c(\alpha_2)c(\beta_2) \end{pmatrix} \tag{A.40}
$$

Propeller 3:

$$
{}^{B}R_{P3}.T_{D3}=\begin{pmatrix} C_d\omega_3^2\left[c((3-1)\frac{2\pi}{6})s(\beta_3)+s((3-1)\frac{2\pi}{6})s(\alpha_3)c(\beta_3)\right] \\ C_d\omega_3^2\left[s((3-1)\frac{2\pi}{6})s(\beta_3)-c((3-1)\frac{2\pi}{6})s(\alpha_3)c(\beta_3)\right] \\ C_d\omega_3^2 c(\alpha_3)c(\beta_3) \end{pmatrix}=C_d\omega_3^2\begin{pmatrix} -\frac{1}{2}s(\beta_3)+\frac{\sqrt{3}}{2}s(\alpha_3)c(\beta_3) \\ \frac{\sqrt{3}}{2}s(\beta_3)+\frac{1}{2}s(\alpha_3)c(\beta_3) \\ c(\alpha_3)c(\beta_3) \end{pmatrix} \tag{A.41}
$$

Propeller 4:

$$
{}^{B}R_{P4}.T_{D4}=\begin{pmatrix} C_d\omega_4^2\left[c((4-1)\frac{2\pi}{6})s(\beta_4)+s((4-1)\frac{2\pi}{6})s(\alpha_4)c(\beta_4)\right] \\ C_d\omega_4^2\left[s((4-1)\frac{2\pi}{6})s(\beta_4)-c((4-1)\frac{2\pi}{6})s(\alpha_4)c(\beta_4)\right] \\ C_d\omega_4^2 c(\alpha_4)c(\beta_4) \end{pmatrix}=C_d\omega_4^2\begin{pmatrix} -s(\beta_4) \\ s(\alpha_4)c(\beta_4) \\ c(\alpha_4)c(\beta_4) \end{pmatrix} \tag{A.42}
$$

Propeller 5:

$$
{}^{B}R_{P5}.T_{D5}=\begin{pmatrix} C_d\omega_5^2\left[c((5-1)\frac{2\pi}{6})s(\beta_5)+s((5-1)\frac{2\pi}{6})s(\alpha_5)c(\beta_5)\right] \\ C_d\omega_5^2\left[s((5-1)\frac{2\pi}{6})s(\beta_5)-c((5-1)\frac{2\pi}{6})s(\alpha_5)c(\beta_5)\right] \\ C_d\omega_5^2 c(\alpha_5)c(\beta_5) \end{pmatrix}=C_d\omega_5^2\begin{pmatrix} -\frac{1}{2}s(\beta_5)-\frac{\sqrt{3}}{2}s(\alpha_5)c(\beta_5) \\ -\frac{\sqrt{3}}{2}s(\beta_5)+\frac{1}{2}s(\alpha_5)c(\beta_5) \\ c(\alpha_5)c(\beta_5) \end{pmatrix} \tag{A.43}
$$

Propeller 6:

$$
{}^{B}R_{P6}.T_{D6}=\begin{pmatrix} C_d\omega_6^2\left[c((6-1)\frac{2\pi}{6})s(\beta_6)+s((6-1)\frac{2\pi}{6})s(\alpha_6)c(\beta_6)\right] \\ C_d\omega_6^2\left[s((6-1)\frac{2\pi}{6})s(\beta_6)-c((6-1)\frac{2\pi}{6})s(\alpha_6)c(\beta_6)\right] \\ C_d\omega_6^2 c(\alpha_6)c(\beta_6) \end{pmatrix}=C_d\omega_6^2\begin{pmatrix} \frac{1}{2}s(\beta_6)-\frac{\sqrt{3}}{2}s(\alpha_6)c(\beta_6) \\ -\frac{\sqrt{3}}{2}s(\beta_6)-\frac{1}{2}s(\alpha_6)c(\beta_6) \\ c(\alpha_6)c(\beta_6) \end{pmatrix} \tag{A.44}
$$

Thrust Component transferred to body frame:

The torque generated in the propeller is transformed into the body frame using the following equation

$$
{}^{B}0_{Pi}\times{}^{B}R_{Pi}.T_{Pi}=\begin{pmatrix} L_{xi}\cos((i-1)\frac{2\pi}{n}) \\ L_{xi}\sin((i-1)\frac{2\pi}{n}) \\ 0 \end{pmatrix}\times\begin{pmatrix} C_l\omega_i^2\left[c((i-1)\frac{2\pi}{n})s(\beta_i)+s((i-1)\frac{2\pi}{n})s(\alpha_i)c(\beta_i)\right] \\ C_l\omega_i^2\left[s((i-1)\frac{2\pi}{n})s(\beta_i)-c((i-1)\frac{2\pi}{n})s(\alpha_i)c(\beta_i)\right] \\ C_l\omega_i^2 c(\alpha_i)c(\beta_i) \end{pmatrix} \tag{A.45}
$$

Therefore for the individual propeller

Propeller 1:

$$
{}^{B}0_{P1}\times{}^{B}R_{P1}.T_{P1}=\begin{pmatrix} L_{x1} \\ 0 \\ 0 \end{pmatrix}\times C_l\omega_1^2\begin{pmatrix} s(\beta_1) \\ -s(\alpha_1)c(\beta_1) \\ c(\alpha_1)c(\beta_1) \end{pmatrix} \tag{A.46}
$$

$$
=C_l\omega_1^2\begin{pmatrix} 0 \\ -L_{x1}c(\alpha_1)c(\beta_1) \\ -L_{x1}s(\alpha_1)c(\beta_1) \end{pmatrix} \tag{A.47}
$$

Propeller 2:

$$
{}^{B}O_{P2} \times {}^{B}R_{P2}.T_{P2} = \begin{pmatrix} \frac{1}{2}L_{x2} \\ \frac{\sqrt{3}}{2}L_{x2} \\ 0 \end{pmatrix} \times C_l\omega_2^2 \begin{pmatrix} \frac{1}{2}s(\beta_2) + \frac{\sqrt{3}}{2}s(\alpha_2)c(\beta_2) \\ \frac{\sqrt{3}}{2}s(\beta_2) - \frac{1}{2}s(\alpha_2)c(\beta_2) \\ c(\alpha_2)c(\beta_2) \end{pmatrix} \tag{A.48}
$$

$$
= C_l\omega_2^2 \begin{pmatrix} \frac{\sqrt{3}}{2}L_{x2}c(\alpha_2)c(\beta_2) \\ -\frac{1}{2}L_{x2}c(\alpha_2)c(\beta_2) \\ \frac{1}{2}L_{x2}\left[\frac{\sqrt{3}}{2}s(\beta_2) - \frac{1}{2}s(\alpha_2)c(\beta_2)\right] - \frac{\sqrt{3}}{2}L_{x2}\left[\frac{1}{2}s(\beta_2) + \frac{\sqrt{3}}{2}s(\alpha_2)c(\beta_2)\right] \end{pmatrix} \tag{A.49}
$$

Propeller 3:

$$
{}^{B}O_{P3} \times {}^{B}R_{P3}.T_{P3} = \begin{pmatrix} -\frac{1}{2}L_{x3} \\ \frac{\sqrt{3}}{2}L_{x3} \\ 0 \end{pmatrix} \times C_l\omega_3^2 \begin{pmatrix} -\frac{1}{2}s(\beta_3) + \frac{\sqrt{3}}{2}s(\alpha_3)c(\beta_3) \\ \frac{\sqrt{3}}{2}s(\beta_3) + \frac{1}{2}s(\alpha_3)c(\beta_3) \\ c(\alpha_3)c(\beta_3) \end{pmatrix} \tag{A.50}
$$

$$
= C_l\omega_3^2 \begin{pmatrix} \frac{\sqrt{3}}{2}L_{x3}c(\alpha_3)c(\beta_3) \\ \frac{1}{2}L_{x3}c(\alpha_3)c(\beta_3) \\ -\frac{1}{2}L_{x3}\left[\frac{\sqrt{3}}{2}s(\beta_3) + \frac{1}{2}s(\alpha_3)c(\beta_3)\right] - \frac{\sqrt{3}}{2}L_{x3}\left[-\frac{1}{2}s(\beta_3) + \frac{\sqrt{3}}{2}s(\alpha_3)c(\beta_3)\right] \end{pmatrix} \tag{A.51}
$$

Propeller 4:

$$
{}^{B}O_{P4} \times {}^{B}R_{P4}.T_{P4} = \begin{pmatrix} -L_{x1} \\ 0 \\ 0 \end{pmatrix} \times C_l\omega_4^2 \begin{pmatrix} -s(\beta_4) \\ s(\alpha_4)c(\beta_4) \\ c(\alpha_4)c(\beta_4) \end{pmatrix} \tag{A.52}
$$

$$
= C_l\omega_4^2 \begin{pmatrix} 0 \\ L_{x1}c(\alpha_4)c(\beta_4) \\ -L_{x1}s(\alpha_4)c(\beta_4) \end{pmatrix} \tag{A.53}
$$

Propeller 5:

$$
{}^{B}O_{P5} \times {}^{B}R_{P5}.T_{P5} = \begin{pmatrix} -\frac{1}{2}L_{x2} \\ -\frac{\sqrt{3}}{2}L_{x2} \\ 0 \end{pmatrix} \times C_l\omega_5^2 \begin{pmatrix} -\frac{1}{2}s(\beta_5) - \frac{\sqrt{3}}{2}s(\alpha_5)c(\beta_5) \\ -\frac{\sqrt{3}}{2}s(\beta_5) + \frac{1}{2}s(\alpha_5)c(\beta_5) \\ c(\alpha_5)c(\beta_5) \end{pmatrix} \tag{A.54}
$$

$$
= C_l\omega_5^2 \begin{pmatrix} -\frac{\sqrt{3}}{2}L_{x2}c(\alpha_5)c(\beta_5) \\ \frac{1}{2}L_{x2}c(\alpha_5)c(\beta_5) \\ \frac{1}{2}L_{x2}\left[\frac{\sqrt{3}}{2}s(\beta_5) - \frac{1}{2}s(\alpha_5)c(\beta_5)\right] - \frac{\sqrt{3}}{2}L_{x2}\left[\frac{1}{2}s(\beta_5) + \frac{\sqrt{3}}{2}s(\alpha_5)c(\beta_5)\right] \end{pmatrix} \tag{A.55}
$$

Propeller 6:

$$
{}^{B}O_{P6} \times {}^{B}R_{P6}.T_{P6} = \begin{pmatrix} \frac{1}{2}L_{x3} \\ -\frac{\sqrt{3}}{2}L_{x3} \\ 0 \end{pmatrix} \times C_l\omega_6^2 \begin{pmatrix} \frac{1}{2}s(\beta_6) - \frac{\sqrt{3}}{2}s(\alpha_6)c(\beta_6) \\ -\frac{\sqrt{3}}{2}s(\beta_6) - \frac{1}{2}s(\alpha_6)c(\beta_6) \\ c(\alpha_6)c(\beta_6) \end{pmatrix} \tag{A.56}
$$

$$
= C_l\omega_6^2 \begin{pmatrix} -\frac{\sqrt{3}}{2}L_{x3}c(\alpha_6)c(\beta_6) \\ -\frac{1}{2}L_{x3}c(\alpha_6)c(\beta_6) \\ -\frac{1}{2}L_{x3}\left[\frac{\sqrt{3}}{2}s(\beta_6) + \frac{1}{2}s(\alpha_6)c(\beta_6)\right] - \frac{\sqrt{3}}{2}L_{x3}\left[-\frac{1}{2}s(\beta_6) + \frac{\sqrt{3}}{2}s(\alpha_6)c(\beta_6)\right] \end{pmatrix} \tag{A.57}
$$

Total torque generated in each propeller:

The torque generated by the individual propeller transfered to the body frame is given by

Propeller 1:

$$
{}^{B}O_{P1} \times {}^{B}R_{P1}.T_{P1} + {}^{B}R_{P1}.T_{D1} = C_l\omega_1^2 \begin{pmatrix} 0 \\ -L_{x1}c(\alpha_1)c(\beta_1) \\ -L_{x1}s(\alpha_1)c(\beta_1) \end{pmatrix} + C_d\omega_1^2 \begin{pmatrix} s(\beta_1) \\ -s(\alpha_1)c(\beta_1) \\ c(\alpha_1)c(\beta_1) \end{pmatrix} \tag{A.58}
$$

Propeller 2:

$^BO_{P2} \times {}^BR_{P2}.T_{P2} + {}^BR_{P2}.T_{D2} =$

$$C_l\omega_2^2 \begin{pmatrix} \frac{\sqrt{3}}{2}L_{x2}c(\alpha_2)c(\beta_2) \\ -\frac{1}{2}L_{x2}c(\alpha_2)c(\beta_2) \\ \frac{1}{2}L_{x2}\left[\frac{\sqrt{3}}{2}s(\beta_2) - \frac{1}{2}s(\alpha_2)c(\beta_2)\right] - \frac{\sqrt{3}}{2}L_{x2}\left[\frac{1}{2}s(\beta_2) + \frac{\sqrt{3}}{2}s(\alpha_2)c(\beta_2)\right] \end{pmatrix} + C_d\omega_2^2 \begin{pmatrix} \frac{1}{2}s(\beta_2) + \frac{\sqrt{3}}{2}s(\alpha_2)c(\beta_2) \\ \frac{\sqrt{3}}{2}s(\beta_2) - \frac{1}{2}s(\alpha_2)c(\beta_2) \\ c(\alpha_2)c(\beta_2) \end{pmatrix} \quad (A.59)$$

Propeller 3:
$^BO_{P3} \times {}^BR_{P3}.T_{P3} + {}^BR_{P3}.T_{D3} =$

$$C_l\omega_3^2 \begin{pmatrix} \frac{\sqrt{3}}{2}L_{x3}c(\alpha_3)c(\beta_3) \\ \frac{1}{2}L_{x3}c(\alpha_3)c(\beta_3) \\ -\frac{1}{2}L_{x3}\left[\frac{\sqrt{3}}{2}s(\beta_3) + \frac{1}{2}s(\alpha_3)c(\beta_3)\right] - \frac{\sqrt{3}}{2}L_{x3}\left[-\frac{1}{2}s(\beta_3) + \frac{\sqrt{3}}{2}s(\alpha_3)c(\beta_3)\right] \end{pmatrix} + C_d\omega_3^2 \begin{pmatrix} -\frac{1}{2}s(\beta_3) + \frac{\sqrt{3}}{2}s(\alpha_3)c(\beta_3) \\ \frac{\sqrt{3}}{2}s(\beta_3) + \frac{1}{2}s(\alpha_3)c(\beta_3) \\ c(\alpha_3)c(\beta_3) \end{pmatrix} \quad (A.60)$$

Propeller 4:

$$^BO_{P4} \times {}^BR_{P4}.T_{P4} + {}^BR_{P4}.T_{D4} = C_l\omega_4^2 \begin{pmatrix} 0 \\ L_{x1}c(\alpha_4)c(\beta_4) \\ -L_{x1}s(\alpha_4)c(\beta_4) \end{pmatrix} + C_d\omega_4^2 \begin{pmatrix} -s(\beta_4) \\ s(\alpha_4)c(\beta_4) \\ c(\alpha_4)c(\beta_4) \end{pmatrix} \quad (A.61)$$

Propeller 5:
$^BO_{P5} \times {}^BR_{P5}.T_{P5} + {}^BR_{P5}.T_{D5} =$

$$C_l\omega_5^2 \begin{pmatrix} -\frac{\sqrt{3}}{2}L_{x2}c(\alpha_5)c(\beta_5) \\ \frac{1}{2}L_{x2}c(\alpha_5)c(\beta_5) \\ \frac{1}{2}L_{x2}\left[\frac{\sqrt{3}}{2}s(\beta_5) - \frac{1}{2}s(\alpha_5)c(\beta_5)\right] - \frac{\sqrt{3}}{2}L_{x2}\left[\frac{1}{2}s(\beta_5) + \frac{\sqrt{3}}{2}s(\alpha_5)c(\beta_5)\right] \end{pmatrix} + C_d\omega_5^2 \begin{pmatrix} -\frac{1}{2}s(\beta_5) - \frac{\sqrt{3}}{2}s(\alpha_5)c(\beta_5) \\ -\frac{\sqrt{3}}{2}s(\beta_5) + \frac{1}{2}s(\alpha_5)c(\beta_5) \\ c(\alpha_5)c(\beta_5) \end{pmatrix} \quad (A.62)$$

Propeller 6:
$^BO_{P6} \times {}^BR_{P6}.T_{P6} + {}^BR_{P6}.T_{D6} =$

$$C_l\omega_6^2 \begin{pmatrix} -\frac{\sqrt{3}}{2}L_{x3}c(\alpha_6)c(\beta_6) \\ -\frac{1}{2}L_{x3}c(\alpha_6)c(\beta_6) \\ -\frac{1}{2}L_{x3}\left[\frac{\sqrt{3}}{2}s(\beta_6) + \frac{1}{2}s(\alpha_6)c(\beta_6)\right] - \frac{\sqrt{3}}{2}L_{x3}\left[-\frac{1}{2}s(\beta_6) + \frac{\sqrt{3}}{2}s(\alpha_6)c(\beta_6)\right] \end{pmatrix} + C_d\omega_6^2 \begin{pmatrix} \frac{1}{2}s(\beta_6) - \frac{\sqrt{3}}{2}s(\alpha_6)c(\beta_6) \\ -\frac{\sqrt{3}}{2}s(\beta_6) - \frac{1}{2}s(\alpha_6)c(\beta_6) \\ c(\alpha_6)c(\beta_6) \end{pmatrix} \quad (A.63)$$

Therefore the torque generated on the body $\boldsymbol{\tau}_h$ is given by,

$$\boldsymbol{\tau}_h = \sum_{i=1}^{n} \left({}^BO_{Pi} \times {}^BR_{Pi}.T_{Pi} + {}^BR_{Pi}.T_{Di}\right) \quad (A.64)$$

Here each column of $\boldsymbol{\tau}_h$ are given by equations from (A.58)-(A.63). This rotational dynamics can be related to equation (5.10) where each individual component of $\boldsymbol{H}(\boldsymbol{\alpha}_h, \boldsymbol{\beta}_h, \boldsymbol{\lambda}_h, \boldsymbol{L}_{x_h}) \in \mathbb{R}^{3\times 6}$ is derived above.

Appendix B

Mechanical Schematics

Figure B.1: CAD model of human-UAV physical interaction setup.

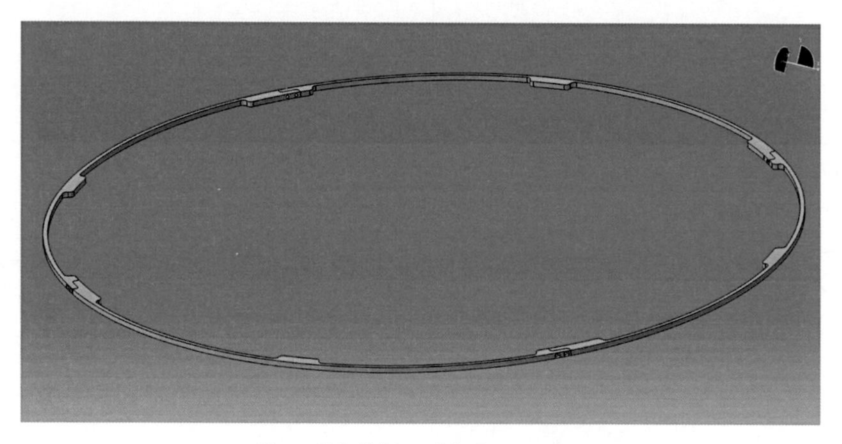

Figure B.2: CAD model of sensor-ring.

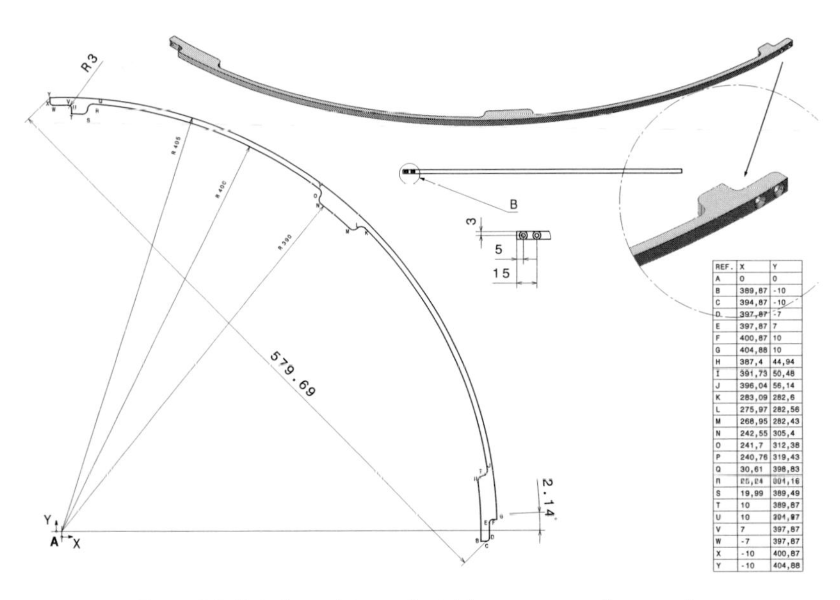

REF.	X	Y
A	0	0
B	389,87	-10
C	394,87	-10
D	397,87	-7
E	397,87	7
F	400,87	10
G	404,88	10
H	387,4	44,94
I	391,73	50,48
J	396,04	56,14
K	283,09	282,6
L	275,97	282,56
M	268,95	282,43
N	242,55	305,4
O	241,7	312,38
P	240,76	319,43
Q	30,61	398,83
R	25,24	[illegible]
S	19,99	389,49
T	10	389,87
U	10	394,87
V	7	397,87
W	-7	397,87
X	-10	400,87
Y	-10	404,88

Figure B.3: Datasheet of sensor-ring. All measurements in *mm* scale.

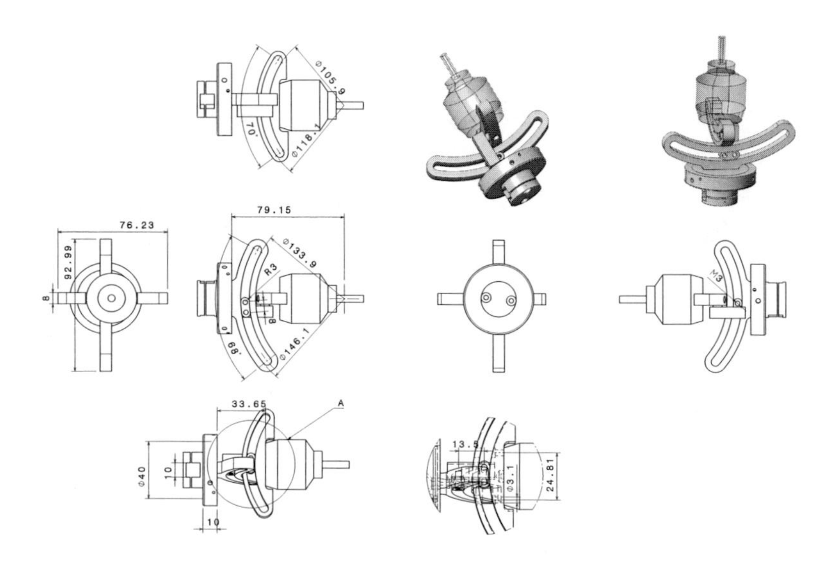

Figure B.4: Datasheet of tilted propeller adaptor. All measurements in *mm* scale.

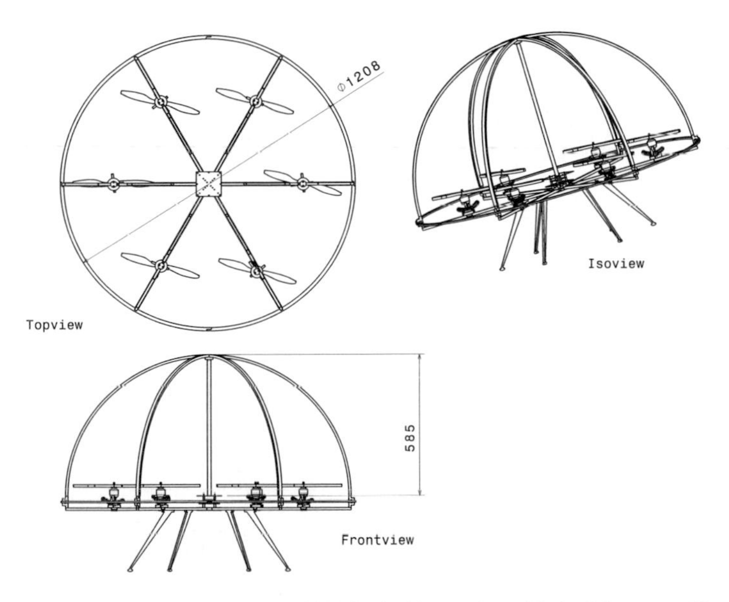

Figure B.5: Datasheet of Human-UAV physical interaction with the fully actuated hexarotor. All measurements in *mm* scale.

Nomenclature

$\dot{x}$	First derivative of x with respect to time
$\ddot{x}$	Second derivative of x with respect to time
$\widehat{x}$	Estimated value of x
$\vec{x}$	Vector representing direction along x
$S(x)$	Skew-symmetric matrix of x
Δx	Difference from the equilibrium value of x
$\|\|x\|\|$	Euclidean norm of vector x
$\int x\,dt$	Integration of the variable x with respect to time
x_h	Variable x is representing a hexarotor parameter
$A(s)/B(s)$	Transfer function of Laplace transforms of input $b(t)$ to output $a(t)$
$[\cdot]_\wedge$	The hat operator from $\mathbb{R}^3 \rightarrow so(3)$
$[\cdot]_\vee$	The inverse (vee) operator from $so(3) \rightarrow \mathbb{R}^3$

Symbols

The most important and common symbols used in this dissertation are listed here.

m	Mass of the quadrotor
m_h	Mass of the hexarotor
g	Acceleration due to gravity
$\boldsymbol{I}_B$	Inertial matrix of quadrotor
$\boldsymbol{I}_{B_h}$	Inertial matrix of hexarotor
b	Propeller lift coefficient
d	Propeller drag coefficient
l	Length of the quadrotor arm
$L_{x_{hi}}$	Arm length of hexarotor
$\mathcal{F}_W$	World inertial frame
$\mathcal{F}_B$	Quadrotor body frame
$\mathcal{F}_H$	Quadrotor horizontal frame
$\mathcal{F}_{B_h}$	Hexarotor body frame
$\mathcal{F}_{P_{hi}}$	Frame associated with the i-th propeller of hexarotor
$\boldsymbol{p}_W$	Position of quadrotor in $\mathcal{F}_W$
$\boldsymbol{p}_d$	Desired trajectory of the quadrotor
$\boldsymbol{p}_a$	Reference or admittance trajectory given to the quadrotor
$\boldsymbol{p}_h$	Position of hexarotor in $\mathcal{F}_W$
$\boldsymbol{p}_{h_d}$	Desired trajectory of the hexarotor
$\boldsymbol{p}_{h_a}$	Reference or admittance trajectory given to the hexarotor
$\boldsymbol{\Theta}_W$	Orientation of quadrotor in $\mathcal{F}_W$
$\boldsymbol{\Theta}_d$	Desired orientation of quadrotor
$\boldsymbol{\Theta}_h$	Orientation of hexarotor in $\mathcal{F}_W$
$\boldsymbol{\Theta}_{h_d}$	Desired orientation of hexarotor
$\boldsymbol{\omega}_B$	Angular velocity of quadrotor in $\mathcal{F}_B$
$\boldsymbol{\omega}_{B_h}$	Angular velocity of hexarotor in $\mathcal{F}_{B_h}$
$\boldsymbol{\xi}_W$	Quadrotor system states
$\boldsymbol{\zeta}$	Generalized velocity vector states
Ω_i	Rotational velocity of i-th propeller
$\boldsymbol{R}_B^W$	Rotation matrix between $\mathcal{F}_W$ and $\mathcal{F}_B$
$\boldsymbol{R}_H^W$	Rotation matrix between $\mathcal{F}_W$ and $\mathcal{F}_H$
$\boldsymbol{R}_B^H$	Rotation matrix between $\mathcal{F}_H$ and $\mathcal{F}_B$
$\boldsymbol{T}(\boldsymbol{\Theta}_W)$	Standard transformation matrix from $\boldsymbol{\omega}_B$ to Euler angle rates $\dot{\boldsymbol{\Theta}}_W$

${}^{W}\boldsymbol{R}_{B_h}$	Rotation matrix representing the orientation of $\mathcal{F}_{B_h}$ w.r.t. $\mathcal{F}_W$ in hexarotor
$\boldsymbol{\tau}$	Torque around the three axes $\vec{\boldsymbol{X}}_B$, $\vec{\boldsymbol{Y}}_B$, $\vec{\boldsymbol{Z}}_B$ in quadrotor
$\boldsymbol{\tau}_h$	Torque around the three axes $\vec{\boldsymbol{X}}_{B_h}$, $\vec{\boldsymbol{Y}}_{B_h}$, $\vec{\boldsymbol{Z}}_{B_h}$ in hexarotor
ρ	Thrust along $\vec{\boldsymbol{Z}}_B$
$\boldsymbol{F}_{ext}$	External forces acting on the quadrotor
$\boldsymbol{F}_a$	Admittance force acting on the quadrotor
$\boldsymbol{\tau}_{ext}$	External torque acting on the quadrotor
$\boldsymbol{\Lambda}_{ext}$	External wrench acting on the quadrotor
$\boldsymbol{\Lambda}_{dis}$	Disturbance wrench acting on the quadrotor
$\boldsymbol{\Lambda}_{int}$	Interaction wrench acting on the quadrotor
$\boldsymbol{r}$	Residual vector
$\boldsymbol{u}$	Control input of quadrotor
$\boldsymbol{u}_h$	Control input of hexarotor
$\boldsymbol{v}$	Virtual control input
$\boldsymbol{\sigma}$	Sliding variable vector
$\boldsymbol{\alpha}_h$	Hexarotor propeller tilt w.r.t. $\vec{\boldsymbol{X}}_{p_h}$
$\boldsymbol{\beta}_h$	Hexarotor propeller tilt w.r.t. $\vec{\boldsymbol{Y}}_{p_h}$

Abbreviations

3D	3 Dimension
6D	6 Dimension
ASTC	Adaptive Super Twisting Controller
BL-CTRL	BrushLess ConTRoL
CAD	Computer Aided Design
CoM	Center of Mass
DoF	Degrees of Freedom
EUROC	European Robotics Challenge
FF	FeedForward
GPIO	General Purpose Input / Output
HIL	Hardware-In-Loop
HRI	Human Robot Interaction
HRPI	Human Robot Physical Interaction
IMU	Inertial Measurement Unit
LED	Light Emitting Diodes
MAV	Micro Aerial Vehicle
MPI	Max Planck Institute
PoC	Point of Contact
QP	Quadratic Programming
ROS	Robot Operating System
STC	Super Twisting Controller
UAV	Unmanned Aerial Vehicle
VICON	Motion Capture System

Bibliography

Abeywardena, D., Kodagoda, S., Dissanayake, G., and Munasinghe, R. (2013). Improved state estimation in quadrotor mavs: A novel drift-free velocity estimator. *IEEE Robotics Automation Magazine*, **20**(4), 32–39.

Albers, A., Trautmann, S., Howard, T., Nguyen, T. A., Frietsch, M., and Sauter, C. (2010). Semi-autonomous flying robot for physical interaction with environment. In *2010 IEEE Conference on Robotics, Automation and Mechatronics*, pages 441–446.

Alexis, K., Nikolakopoulos, G., and Tzes, A. (2011). Switching model predictive attitude control for a quadrotor helicopter subject to atmospheric disturbances. *Control Engineering Practice*, **19**(10), 1195–1207. cited By 83.

Antonelli, G., Arrichiello, F., Chiaverini, S., and Giordano, P. R. (2013). Adaptive trajectory tracking for quadrotor mavs in presence of parameter uncertainties and external disturbances. In *2013 IEEE/ASME International Conference on Advanced Intelligent Mechatronics*, pages 1337–1342.

Augugliaro, F. and D'Andrea, R. (2013). Admittance control for physical human-quadrocopter interaction. In *Control Conference (ECC), 2013 European*, pages 1805–1810.

Augugliaro, F., Lupashin, S., Hamer, M., Male, C., Hehn, M., Mueller, M. W., Willmann, J. S., Gramazio, F., Kohler, M., and D'Andrea, R. (2014). The flight assembled architecture installation: Cooperative construction with flying machines. *IEEE Control Systems*, **34**(4), 46–64.

Bellens, S., Schutter, J. D., and Bruyninckx, H. (2012). A hybrid pose / wrench control framework for quadrotor helicopters. In *Robotics and Automation (ICRA), 2012 IEEE International Conference on*, pages 2269–2274.

Bouabdallah, S. and Siegwart, R. (2005). Backstepping and sliding-mode techniques applied to an indoor micro quadrotor. In *Proceedings of the 2005 IEEE International Conference on Robotics and Automation*, pages 2247–2252.

Bouabdallah, S. and Siegwart, R. (2007). Full control of a quadrotor. In *2007 IEEE/RSJ International Conference on Intelligent Robots and Systems*, pages 153–158.

Bouadi, H., Simoes Cunha, S., Drouin, A., and Mora-Camino, F. (2011). Adaptive sliding mode control for quadrotor attitude stabilization and altitude tracking. In *Computational Intelligence and Informatics (CINTI), 2011 IEEE 12th International Symposium on*, pages 449–455.

Bouktir, Y., Haddad, M., and Chettibi, T. (2008). Trajectory planning for a quadrotor helicopter. In *2008 16th Mediterranean Conference on Control and Automation*, pages 1258–1263.

Bresciani, T. (2008). Modelling, identification and control of a quadrotor helicopter. MSc Theses, Lund University.

Brescianini, D. and D'Andrea, R. (2016). Design, modeling and control of an omni-directional aerial vehicle. In *2016 IEEE International Conference on Robotics and Automation (ICRA)*, pages 3261–3266.

Briod, A., Kornatowski, P., Zufferey, J.-C., and Floreano, D. (2014). A collision-resilient flying robot. *J. Field Robot.*, **31**(4), 496–509.

Cai, G., Chen, B. M., Lee, T. H., and Dong, M. (2008). Design and implementation of a hardware-in-the-loop simulation system for small-scale uav helicopters. In *2008 IEEE International Conference on Automation and Logistics*, pages 29–34.

Castaneda, H., Salas-Pena, O., and de Leon Morales, J. (2013). Adaptive super twisting flight control-observer for a fixed wing uav. In *Unmanned Aircraft Systems (ICUAS), 2013 International Conference on*, pages 1004–1013.

Cauchard, J. R., E, J. L., Zhai, K. Y., and Landay, J. A. (2015). Drone & me: An exploration into natural human-drone interaction. In *Proceedings of the 2015 ACM International Joint Conference on Pervasive and Ubiquitous Computing*, UbiComp '15, pages 361–365, New York, NY, USA. ACM.

Chandhrasekaran, V. K. and Choi, E. (2010). Fault tolerance system for uav using hardware in the loop simulation. In *New Trends in Information Science and Service Science (NISS), 2010 4th International Conference on*, pages 293–300.

Chen, W.-H., Ballance, D. J., Gawthrop, P. J., and O'Reilly, J. (2000). A nonlinear disturbance observer for robotic manipulators. *IEEE Transactions on Industrial Electronics*, **47**(4), 932–938.

Cheon, Y., Lee, D., Lee, I.-B., and Sung, S. W. (2013). A new pid auto-tuning strategy with operational optimization for mcfc systems. In *Control Conference (ASCC), 2013 9th Asian*, pages 1–6.

Darivianakis, G., Alexis, K., Burri, M., and Siegwart, R. (2014). Hybrid predictive control for aerial robotic physical interaction towards inspection operations. In *2014 IEEE International Conference on Robotics and Automation (ICRA)*, pages 53–58.

De Luca, A. and Mattone, R. (2003). Actuator failure detection and isolation using generalized momenta. In *Robotics and Automation, 2003. Proceedings. ICRA '03. IEEE International Conference on*, volume 1, pages 634–639 vol.1.

De Luca, A. and Oriolo, G. (2002). Trajectory planning and control for planar robots with passive last joint. *Intern. Journal of Robotics Research*, **21**(5-6), 575–590.

Derafa, L., Fridman, L., Benallegue, A., and Ouldali, A. (2010). Super twisting control algorithm for the four rotors helicopter attitude tracking problem. In *Variable Structure Systems (VSS), 2010 11th International Workshop on*, pages 62–67.

Derafa, L., Benallegue, A., and Fridman, L. (2012). Super twisting control algorithm for the attitude tracking of a four rotors {UAV}. *Journal of the Franklin Institute*, **349**(2), 685 – 699. Advances in Guidance and Control of Aerospace Vehicles using Sliding Mode Control and Observation Techniques.

Erginer, B. and Altug, E. (2007). Modeling and pd control of a quadrotor vtol vehicle. In *2007 IEEE Intelligent Vehicles Symposium*, pages 894–899.

Erlic, M. and Lu, W. S. (1993). A reduced-order adaptive velocity observer for manipulator control. In *[1993] Proceedings IEEE International Conference on Robotics and Automation*, pages 328–333 vol.2.

Flores, G. R., Escareño, J., Lozano, R., and Salazar, S. (2011). Quad-Tilting Rotor Convertible MAV: Modeling and Real-Time Hover Flight Control. *Journal of Intelligent & Robotic Systems*, **65**(1-4), 457–471.

Fraundorfer, F., Heng, L., Honegger, D., Lee, G. H., Meier, L., Tanskanen, P., and Pollefeys, M. (2012). Vision-based autonomous mapping and exploration using a quadrotor mav. In *2012 IEEE/RSJ International Conference on Intelligent Robots and Systems*, pages 4557–4564.

Fumagalli, M. and Carloni, R. (2013). A modified impedance control for physical interaction of uavs. In *2013 IEEE/RSJ International Conference on Intelligent Robots and Systems*, pages 1979–1984.

Fumagalli, M., Naldi, R., Macchelli, A., Carloni, R., Stramigioli, S., and Marconi, L. (2012). Modeling and control of a flying robot for contact inspection. In *2012 IEEE/RSJ International Conference on Intelligent Robots and Systems*, pages 3532–3537.

Fumagalli, M., Naldi, R., Macchelli, A., Forte, F., Keemink, A. Q. L., Stramigioli, S., Carloni, R., and Marconi, L. (2014). Developing an aerial manipulator prototype: Physical interaction with the environment. *IEEE Robotics Automation Magazine*, **21**(3), 41–50.

Gioioso, G., Franchi, A., Salvietti, G., Scheggi, S., and Prattichizzo, D. (2014a). The Flying Hand: a formation of uavs for cooperative aerial tele-manipulation. In *2014 IEEE Int. Conf. on Robotics and Automation*, pages 4335–4341, Hong Kong, China.

Gioioso, G., Ryll, M., Prattichizzo, D., Blthoff, H. H., and Franchi, A. (2014b). Turning a near-hovering controlled quadrotor into a 3d force effector. In *2014 IEEE International Conference on Robotics and Automation (ICRA)*, pages 6278–6284.

Grabe, V., Riedel, M., Bülthoff, H. H., Giordano, P. R., and Franchi, A. (2013). The telekyb framework for a modular and extendible ros-based quadrotor control. In *Mobile Robots (ECMR), 2013 European Conference on*, pages 19–25.

Guenard, N., Hamel, T., and Mahony, R. (2008). A practical visual servo control for an unmanned aerial vehicle. *IEEE Transactions on Robotics*, **24**(2), 331–340.

Hacksel, P. and Salcudean, S. (1994). Estimation of environment forces and rigid-body velocities using observers. In *Robotics and Automation, 1994. Proceedings., 1994 IEEE International Conference on*, pages 931–936 vol.2.

Heng, L., Meier, L., Tanskanen, P., Fraundorfer, F., and Pollefeys, M. (2011). Autonomous obstacle avoidance and maneuvering on a vision-guided mav using on-board processing. In *Robotics and Automation (ICRA), 2011 IEEE International Conference on*, pages 2472–2477.

Isidori, A. (1995). *Nonlinear Control Systems, 3rd edition*. Springer.

Keemink, A., Fumagalli, M., Stramigioli, S., and Carloni, R. (2012). Mechanical design of a manipulation system for unmanned aerial vehicles. *2012 IEEE International Conference on Robotics and Automation*, pages 3147–3152.

Kendoul, F., Fantoni, I., and Lozano, R. (2006). Modeling and control of a small autonomous aircraft having two tilting rotors. *IEEE Trans. on Robotics*, **22**(6), 1297–1302.

Khalil, H. K. (2002). *Nonlinear systems*. Prentice Hall, Upper Saddle River (New Jersey).

Khalil, W. and Dombre, E. (2004). *Modeling, Identification and Control of Robots*. Butterworth-Heinemann, Oxford.

Kim, S., Seo, H., Choi, S., and Kim, H. J. (2016). Vision-guided aerial manipulation using a multirotor with a robotic arm. *IEEE/ASME Transactions on Mechatronics*, **21**(4), 1912–1923.

Lee, D., Jin Kim, H., and Sastry, S. (2009). Feedback linearization vs. adaptive sliding mode control for a quadrotor helicopter. *International Journal of Control, Automation and Systems*, **7**(3), 419–428.

Lee, D., Franchi, A., Son, H. I., Ha, C., Bülthoff, H., and Giordano, P. (2013). Semiautonomous haptic teleoperation control architecture of multiple unmanned aerial vehicles. *Mechatronics, IEEE/ASME Transactions on*, **18**(4), 1334–1345.

Lee, T., Leok, M., and McClamroch, N. H. (2010). Geometric tracking control of a quadrotor UAV on $SE(3)$. In *2010 IEEE Conf. on Decision and Control*, pages 5420–5425.

Levant, A. (1993). Sliding order and sliding accuracy in sliding mode control. *Intern. Journal of Control*, **58**(6), 1247–1263.

Lichtenstern, M., Frassl, M., Perun, B., and Angermann, M. (2012). A prototyping environment for interaction between a human and a robotic multi-agent system. In *Human-Robot Interaction (HRI), 2012 7th ACM/IEEE International Conference on*, pages 185–186.

Lindsey, Q., Mellinger, D., and Kumar, V. (2011). Construction of cubic structures with quadrotor teams. In *2011 Robotics: Science and Systems*, Los Angeles, CA.

Lindsey, Q., Mellinger, D., and Kumar, V. (2012). Construction with quadrotor teams. *Autonomous Robots*, **33**(3), 323–336.

Lippiello, V. and Ruggiero, F. (2012). Exploiting redundancy in cartesian impedance control of uavs equipped with a robotic arm. In *2012 IEEE/RSJ International Conference on Intelligent Robots and Systems*, pages 3768–3773.

Liu, Y., Montenbruck, J. M., Stegagno, P., Allgöwer, F., and Zell, A. (2015). A robust nonlinear controller for nontrivial quadrotor maneuvers: Approach and verification. In *Proc. of the IEEE/RSJ International Conference on Intelligent Robots and Systems*, pages 600–606.

Liu, Y., Rajappa, S., Montenbruck, J. M., Stegagno, P., Bülthoff, H., Allgöwer, F., and Zell, A. (2017a). A robust nonlinear control approach to nontrivial quadrotor maneuvers under disturbances. *IEEE Transactions on Control Systems Technology*, **Submitted**.

Liu, Y., Rajappa, S., Stegagno, P., Allgöwer, F., and Zell, A. (2017b). A robust nonlinear predictive control approach to multirotor maneuvering with obstacle avoidance. *2017 IEEE International Conference on Robotics and Automation (ICRA)*, **Submitted**.

Luca, A. D., Schroder, D., and Thummel, M. (2007). An acceleration-based state observer for robot manipulators with elastic joints. In *Proceedings 2007 IEEE International Conference on Robotics and Automation*, pages 3817–3823.

Luenberger, D. G. (1979). *Introduction to dynamic systems : theory, models, and applications*. J. Wiley & Sons, New York, Chichester, Brisbane.

Lupashin, S., Schllig, A., Sherback, M., and D'Andrea, R. (2010). A simple learning strategy for high-speed quadrocopter multi-flips. In *Robotics and Automation (ICRA), 2010 IEEE International Conference on*, pages 1642–1648.

Madani, T. and Benallegue, A. (2006). Backstepping control for a quadrotor helicopter. In *Intelligent Robots and Systems, 2006 IEEE/RSJ International Conference on*, pages 3255–3260.

Magrini, E., Flacco, F., and Luca, A. D. (2014). Estimation of contact forces using a virtual force sensor. In *2014 IEEE/RSJ International Conference on Intelligent Robots and Systems*, pages 2126–2133.

Marconi, L., Naldi, R., and Gentili, L. (2011). Modeling and control of a flying robot interacting with the environment. *Automatica*, **47**(12), 2571–2583.

Mebarki, R., Lippiello, V., and Siciliano, B. (2014). Image-based control for dynamically cross-coupled aerial manipulation. In *2014 IEEE/RSJ International Conference on Intelligent Robots and Systems*, pages 4827–4833.

Mebarki, R., Lippiello, V., and Siciliano, B. (2015). Nonlinear visual control of unmanned aerial vehicles in gps-denied environments. *IEEE Transactions on Robotics*, **31**(4), 1004–1017.

Mellinger, D., Shomin, M., Michael, N., and Kumar, V. (2010). Cooperative grasping and transport using multiple quadrotors. *International Symposium on Distributed Autonomous Robotic Systems*.

Mellinger, D., Michael, N., and Kumar, V. (2012). Trajectory generation and control for precise aggressive maneuvers with quadrotors. *The International Journal of Robotics Research*, **31**(5), 664–674.

Michael, N., Mellinger, D., Lindsey, Q., and Kumar, V. (2010). The grasp multiple micro-uav testbed. *Robotics Automation Magazine, IEEE*, **17**(3), 56–65.

Mistler, V., Benallegue, A., and M'Sirdi, N. (2001). Exact linearization and noninteracting control of a 4 rotors helicopter via dynamic feedback. In *Robot and Human Interactive Communication, 2001. Proceedings. 10th IEEE International Workshop on*, pages 586–593.

Mokhtari, A. and Benallegue, A. (2004). Dynamic feedback controller of euler angles and wind parameters estimation for a quadrotor unmanned aerial vehicle. In *Robotics and Automation, 2004. Proceedings. ICRA '04. 2004 IEEE International Conference on*, volume 3, pages 2359–2366 Vol.3.

Monajjemi, V. M., Wawerla, J., Vaughan, R., and Mori, G. (2013). Hri in the sky: Creating and commanding teams of uavs with a vision-mediated gestural interface. In *2013 IEEE/RSJ International Conference on Intelligent Robots and Systems*, pages 617–623.

Montufar, D. I., Muoz, F., Espinoza, E. S., Garcia, O., and Salazar, S. (2014). Multi-uav testbed for aerial manipulation applications. In *2014 International Conference on Unmanned Aircraft Systems (ICUAS)*, pages 830–835.

Mueggler, E., Faessler, M., Fontana, F., and Scaramuzza, D. (2014). Aerial-guided navigation of a ground robot among movable obstacles. In *2014 IEEE International Symposium on Safety, Security, and Rescue Robotics (2014)*, pages 1–8.

Nagi, J., Giusti, A., Gambardella, L. M., and Caro, G. A. D. (2014). Human-swarm interaction using spatial gestures. In *2014 IEEE/RSJ International Conference on Intelligent Robots and Systems*, pages 3834–3841.

Naldi, R., Gentili, L., Marconi, L., and Sala, A. (2010). Design and experimental validation of a nonlinear control law for a ducted-fan miniature aerial vehicle. *Control Engineering Practice*, **18**(7), 747–760.

Naseer, T., Sturm, J., and Cremers, D. (2013). Followme: Person following and gesture recognition with a quadrocopter. In *2013 IEEE/RSJ International Conference on Intelligent Robots and Systems*, pages 624–630.

Ng, W. S. and Sharlin, E. (2011). Collocated interaction with flying robots. In *2011 RO-MAN*, pages 143–149.

Nguyen, H. N. and Lee, D. (2013). Hybrid force/motion control and internal dynamics of quadrotors for tool operation. In *2013 IEEE/RSJ International Conference on Intelligent Robots and Systems*, pages 3458–3464.

Odelga, M., Stegagno, P., Bülthoff, H. H., and Ahmad, A. (2015). A setup for multi-uav hardware-in-the-loop simulations. In *2015 Workshop on Research, Education and Development of Unmanned Aerial Systems (RED-UAS)*, pages 204–210.

Odelga, M., Stegagno, P., and Bülthoff, H. H. (2016). A fully actuated quadrotor uav with a propeller tilting mechanism: Modeling and control. In *2016 IEEE International Conference on Advanced Intelligent Mechatronics (AIM)*, pages 306–311.

Oner, K. T., Cetinsoy, E., Unel, M., Aksit, M. F., Kandemir, I., and Gulez, K. (2008). Dynamic model and control of a new quadrotor unmanned aerial vehicle with tilt-wing mechanism. In *Proc. of the 2008 World Academy of Science, Engineering and Technology*, pages 58–63.

Orsag, M., Korpela, C., and Oh, P. (2013). Modeling and control of mm-uav: Mobile manipulating unmanned aerial vehicle. *Journal of Intelligent and Robotic Systems*, **69**(1-4), 227–240.

Palunko, I., Cruz, P., and Fierro, R. (2012). Agile load transportation : Safe and efficient load manipulation with aerial robots. *IEEE Robotics Automation Magazine*, **19**(3), 69–79.

Pfeil, K., Koh, S. L., and LaViola, J. (2013). Exploring 3d gesture metaphors for interaction with unmanned aerial vehicles. In *Proceedings of the 2013 International Conference on Intelligent User Interfaces*, IUI '13, pages 257–266, New York, NY, USA. ACM.

Plestan, F., Shtessel, Y., Brégeault, V., and Poznyak, A. (2010). New methodologies for adaptive sliding mode control. *Intern. Journal of Control*, **83**(9), 1907–1919.

Pounds, P. E. I., Bersak, D. R., and Dollar, A. M. (2011). Grasping from the air: Hovering capture and load stability. In *Robotics and Automation (ICRA), 2011 IEEE International Conference on*, pages 2491–2498.

Quigley, M., Goodrich, M. A., and Beard, R. W. (2004). Semi-autonomous human-uav interfaces for fixed-wing mini-uavs. In *Intelligent Robots and Systems, 2004. (IROS 2004). Proceedings. 2004 IEEE/RSJ International Conference on*, volume 3, pages 2457–2462 vol.3.

Raffo, G. V., Ortega, M. G., and Rubio, F. R. (2010). An integral predictive/nonlinear h-infinity control structure for a quadrotor helicopter. *Automatica*, **46**(1), 29–39.

Rajappa, S., Ryll, M., Bülthoff, H. H., and Franchi, A. (2015). Modeling, control and design optimization for a fully-actuated hexarotor aerial vehicle with tilted propellers. In *2015 IEEE International Conference on Robotics and Automation (ICRA)*, pages 4006–4013.

Rajappa, S., Masone, C., Bülthoff, H. H., and Stegagno, P. (2016). Adaptive super twisting controller for a quadrotor uav. In *2016 IEEE International Conference on Robotics and Automation (ICRA)*, pages 2971–2977.

Rajappa, S., Bülthoff, H., and Stegagno, P. (2017a). Design and implementation of a novel architecture for human-uav physical interaction. *International Journal of Robotics Research (IJRR)*, **In Press**.

Rajappa, S., Bülthoff, H., and Stegagno, P. (2017b). Human-UAV physical interaction with a fully actuated UAV. *2017 IEEE/RSJ International Conference on Intelligent Robots and Systems*, **Submitted**.

Ritz, R., Müller, M. W., Hehn, M., and D'Andrea, R. (2012). Cooperative quadrocopter ball throwing and catching. In *2012 IEEE/RSJ International Conference on Intelligent Robots and Systems*, pages 4972–4978.

Roberts, A. and Tayebi, A. (2011). Adaptive position tracking of vtol uavs. *IEEE Transactions on Robotics*, **27**(1), 129–142.

Ruggiero, F., Cacace, J., Sadeghian, H., and Lippiello, V. (2014). Impedance control of vtol uavs with a momentum-based external generalized forces estimator. In *2014 IEEE International Conference on Robotics and Automation (ICRA)*, pages 2093–2099.

Ryll, M., Bülthoff, H. H., and Giordano, P. R. (2012). Modeling and control of a quadrotor UAV with tilting propellers. *2012 IEEE International Conference on Robotics and Automation*, pages 4606–4613.

Salazar, S., Romero, H., Lozano, R., and Castillo, P. (2008). Modeling and Real-Time Stabilization of an Aircraft Having Eight Rotors. *Journal of Intelligent and Robotic Systems*, **54**(1-3), 455–470.

Sanchez, A., Escareño, J., Garcia, O., and Lozano, R. (2008). Autonomous hovering of a noncyclic tiltrotor UAV: Modeling, control and implementation. In *Proc. of the 17th IFAC Wold Congress*, pages 803–808.

Sanna, A., Lamberti, F., Paravati, G., and Manuri, F. (2013). A kinect-based natural interface for quadrotor control. *Entertainment Computing*, **4**(3), 179 – 186.

Scholten, J. L. J., Fumagalli, M., Stramigioli, S., and Carloni, R. (2013). Interaction control of an uav endowed with a manipulator. In *Robotics and Automation (ICRA), 2013 IEEE International Conference on*, pages 4910–4915.

Shtessel, Y., Edwards, C., Fridman, L., and Levant, A. (2014). *Sliding Mode Control and Observation*. Springer.

Shtessel, Y. B., Moreno, J. A., Plestan, F., Fridman, L., and Poznyak, A. S. (2010). Supertwisting adaptive sliding mode control: A Lyapunov design. In *2010 IEEE Conf. on Decision and Control*, pages 5109–5113.

Shtessel, Y. B., Taleb, M., and Plestan, F. (2012). A novel adaptive-gain supertwisting sliding mode controller: Methodology and application. *Automatica*, **48**(5), 759–769.

Spica, R., Giordano, P. R., Ryll, M., Bülthoff, H. H., and Franchi, A. (2013). An open-source hardware/software architecture for quadrotor uavs. In *2nd IFAC Workshop on Research, Education and Development of Unmanned Aerial Systems*, Compiegne, France.

Stegagno, P., Cognetti, M., Rosa, L., Peliti, P., and Oriolo, G. (2013). Relative localization and identification in a heterogeneous multi-robot system. In *Robotics and Automation (ICRA), 2013 IEEE International Conference on*, pages 1857–1864.

Suarez, A., Heredia, G., and Ollero, A. (2015). Lightweight compliant arm for aerial manipulation. In *2015 IEEE/RSJ International Conference on Intelligent Robots and Systems (IROS)*, pages 1627–1632.

Szafir, D., Mutlu, B., and Fong, T. (2015). Communicating directionality in flying robots. In *Proceedings of the Tenth Annual ACM/IEEE International Conference on Human-Robot Interaction*, HRI '15, pages 19–26, New York, NY, USA. ACM.

Takakura, S., Murakami, T., and Ohnishi, K. (1989). An approach to collision detection and recovery motion in industrial robot. In *Industrial Electronics Society, 1989. IECON '89., 15th Annual Conference of IEEE*, pages 421–426 vol.2.

Tomic, T. and Haddadin, S. (2015). Simultaneous estimation of aerodynamic and contact forces in flying robots: Applications to metric wind estimation and collision detection. In *2015 IEEE International Conference on Robotics and Automation (ICRA)*, pages 5290–5296.

Voos, H. (2009). Nonlinear control of a quadrotor micro-uav using feedback-linearization. In *Mechatronics, 2009. ICM 2009. IEEE International Conference on*, pages 1–6.

Voyles, R. and Jiang, G. (2012). Hexrotor uav platform enabling dextrous interaction with structurespreliminary work. In *2012 IEEE International Symposium on Safety, Security, and Rescue Robotics (SSRR)*, pages 1–7.

Xu, R. and Ozguner, U. (2006). Sliding mode control of a quadrotor helicopter. In *Decision and Control, 2006 45th IEEE Conference on*, pages 4957–4962.

Yüksel, B., Secchi, C., Bülthoff, H., and Franchi, A. (2014). A nonlinear force observer for quadrotors and application to physical interactive tasks. In *Advanced Intelligent Mechatronics (AIM), 2014 IEEE/ASME International Conference on*, pages 433–440.

Yüksel, B., Secchi, C., Bülthoff, H. H., and Franchi, A. (2014). Reshaping the physical properties of a quadrotor through ida-pbc and its application to aerial physical interaction. In *2014 IEEE Int. Conf. on Robotics and Automation*, pages 6258–6265, Hong Kong, China.

Yüksel, B., Mahboubi, S., Secchi, C., Bülthoff, H. H., and Franchi, A. (2015). Design, identification and experimental testing of a light-weight flexible-joint arm for aerial physical interaction. In *2015 IEEE International Conference on Robotics and Automation (ICRA)*, pages 870–876.